Lecture Notes in Computer Science 16511

Founding Editors

Gerhard Goos
Juris Hartmanis

The series Lecture Notes in Computer Science (LNCS), including its subseries Lecture Notes in Artificial Intelligence (LNAI) and Lecture Notes in Bioinformatics (LNBI), has established itself as a medium for the publication of new developments in computer science and information technology research, teaching, and education.

LNCS enjoys close cooperation with the computer science R & D community, the series counts many renowned academics among its volume editors and paper authors, and collaborates with prestigious societies. Its mission is to serve this international community by providing an invaluable service, mainly focused on the publication of conference and workshop proceedings and postproceedings. LNCS commenced publication in 1973.

Camille Fayollas · Pieter Van Gorp ·
Mahmoud Baghdadi · Achim Ebert · Jun Hu ·
Shah Rukh Humayoun · Sapna Jaidka ·
Kris Luyten · Tilo Mentler · Philippe Palanque ·
Parvaneh Parvin · Lucio Davide Spano ·
Simone Stumpf · Gerrit van der Veer ·
Luciana Zaina · Jürgen Ziegler
Editors

Engineering Interactive Computer Systems

EICS 2025 International Workshops

Trier, Germany, June 23–27, 2025
Revised Selected Papers

Editors
Camille Fayollas
Université Toulouse Capitole
Toulouse, France

Mahmoud Baghdadi
University of Kaiserslautern-Landau
Kaiserslautern, Germany

Jun Hu
Eindhoven University of Technology
Eindhoven, The Netherlands

Sapna Jaidka
University of Waikato
Hillcrest, New Zealand

Tilo Mentler
Trier University
Trier, Germany

Parvaneh Parvin
Eindhoven University of Technology
Eindhoven, The Netherlands

Simone Stumpf
University of Glasgow
Glasgow, UK

Luciana Zaina
The Open University
São Paulo, Brazil

Pieter Van Gorp
Eindhoven University of Technology
Eindhoven, The Netherlands

Achim Ebert
University of Kaiserslautern-Landau
Kaiserslautern, Germany

Shah Rukh Humayoun
San Francisco State University
San Francisco, CA, USA

Kris Luyten
Hasselt University
Hasselt, Belgium

Philippe Palanque
Université de Toulouse
Toulouse, France

Lucio Davide Spano
University of Cagliari
Cagliari, Italy

Gerrit van der Veer
Radboud University
Nijmegen, The Netherlands

Jürgen Ziegler
University of Duisburg- Essen
Duisburg, Germany

ISSN 0302-9743 ISSN 1611-3349 (electronic)
Lecture Notes in Computer Science
ISBN 978-3-032-26050-5 ISBN 978-3-032-26051-2 (eBook)
https://doi.org/10.1007/978-3-032-26051-2

Foreword

This volume presents a selection of revised and extended papers originating from the Doctoral Consortium and the Workshops held in conjunction with the 17th ACM SIGCHI Symposium on Engineering Interactive Computing Systems (EICS 2025), which took place in Trier, Germany, in June 2025. While EICS has a long and well-established history as an ACM SIGCHI symposium since 2009, this is only the third time that contributions from EICS workshops have been brought together into a single edited volume.

The Symposium on Engineering Interactive Computing Systems (EICS) is the premier venue for research at the intersection of Human–Computer Interaction (HCI) and Software Engineering. EICS is jointly organized by the ACM Special Interest Group on Computer–Human Interaction (SIGCHI) and the IFIP Working Group on User Interface Engineering (WG 2.7/13.4). Now in its seventeenth edition, EICS is widely recognized for rigorously advancing and disseminating research that carefully balances user interface design, software engineering principles, and computational interaction. The symposium emphasizes models, languages, notations, methods, techniques, and tools that support the design, development, validation, and verification of interactive systems. While this scope encompasses traditional application domains such as aviation and command-and-control systems, EICS also promotes research on emerging topics in the engineering of interactive systems.

All contributions to EICS were subject to a thorough peer-review process. For the Doctoral Consortium, submissions were evaluated based on relevance to EICS, originality, methodological and technical soundness, and clarity of presentation. The Doctoral Consortium was chaired by Jürgen Ziegler (University of Duisburg-Essen), Sapna Jaidka (University of Waikato), and Tilo Mentler (Trier University of Applied Sciences). For the workshops, proposals were reviewed based on relevance to EICS, the scientific background and experience of the organizers, and the quality of dissemination and engagement plans. Selection was overseen by Camille Fayollas (Université Toulouse Capitole/IRIT, France) and Pieter Van Gorp (Eindhoven University of Technology, The Netherlands).

Authors revised their submissions after reviews before the event. Then, discussions and a second round of reviews ensured that only high-quality extended papers were included in these proceedings. The Doctoral Consortium papers were presented on June 23, 2025, and the workshop papers on June 24, 2025.

The editors gratefully acknowledge the constructive contributions of anonymous reviewers from several institutions. In addition, workshop and consortium participants provided valuable feedback through lively discussions during the conference.

The remainder of this foreword presents the highlights of the Doctoral Consortium and the three EICS 2025 workshops.

The **EICS 2025 Doctoral Consortium** offered doctoral researchers a forum to present their research objectives and intermediate results, and to engage in in-depth discussion with senior experts and peers. The contributions reflected the breadth of challenges in engineering interactive computing systems and offered promising perspectives for future research.

One contribution, by Cai, explored interdisciplinary engineering processes at the intersection of human–computer interaction, fashion technology, and environmental sensing. The remaining contributions focused on interaction and interface design. Latreche investigated gestural interaction with cubic tangible objects in educational settings. Sahraoui introduced the concept of Extra-User Interfaces, enabling users to exercise greater control over adaptive user interfaces. Krings addressed cross-reality context awareness and proposed a taxonomy to guide the development of reusable building blocks for cross-reality systems. Collectively, these contributions demonstrated the vitality of current doctoral research in the EICS community and pointed toward impactful future publications.

The **Third Workshop on Engineering Interactive Systems Embedding AI Technologies (EISEAIT 2025)**, organized under **IFIP WG 2.7/13.4**, continued its mission of advancing methods, tools, and conceptual foundations for integrating AI technologies into interactive systems across the engineering lifecycle. This year's contributions reflected the growing maturity of the field, addressing data-to-interaction pipelines, explainability in critical domains, model-driven engineering for human-centered AI, formal verification support, trustworthy automation, and conversational transparency.

Zeppieri et al. proposed a domain-agnostic pipeline that transformed natural-language and document-based inputs into structured data using LLM agents, emphasizing mixed-initiative interaction and human control. Fayollas, Garouani, and Martinie examined the integration of explainable AI in safety-critical systems, grounded in an aviation case study on turbulence detection. Conrardy, Capozucca, and Cabot introduced a unified user modeling language consolidating existing approaches into a single meta-model for personalized, human-centered AI systems. Moreira and Campos studied the potential of LLMs to explain model-checking counterexamples to non-experts, highlighting opportunities for combining formal verification and natural language generation. Thys, Vanacken, and Rovelo Ruiz presented an AutoML pipeline tailored for novice users and derived design principles for trustworthy automation. Finally, Vanbrabant, Rovelo Ruiz, and Vanacken introduced MATCH, a framework for transparent and controllable multi-agent conversational XAI systems.

Together, these contributions strengthened the methodological foundations of engineering interactive systems embedding AI and charted a research trajectory centered on transparency, controllability, and human-centeredness.

The workshop **Experience 2.0 and Beyond – From Visuals to Human-Centered Multimedia**, organized by **IFIP Working Group on Visualization (WG 13.7)**, built on the first edition held at EICS 2024. It provided a platform for exploring innovative approaches to engineering AR and MR applications that span devices and modalities, with the goal of making immersive and multimedia experiences more accessible.

Beese et al. investigated spatial awareness in complex virtual buildings and demonstrated that the presence of landmarks significantly improved users' ability to form accurate cognitive maps. Nass-Bauer, Ebert, and Lachmann examined multimodal interaction in AI-assisted design education, arguing for a rethinking of generative-AI-supported learning as a coordinated, multimodal experience. Klingshirn, Garth, and Ebert analyzed challenges of explainable AI in extended reality, identifying open research questions for developing transparent and trustworthy XR systems. Klein and Wichert introduced the concept of Silent Interfaces for Active Assisted Living, presenting field results that showed improved user acceptance, safety, and autonomy.

The workshop **Interactive AI for Preventive Health: Personalization, Gamification, and Ethics (IAI4PH 2025)** focused on interactive AI systems that support preventive health through personalization, engagement, and responsible design. The selected contributions spanned social robotics, conversational agents, gamified rehabilitation, and public-health perspectives.

Li et al. presented a robot-supported conversational platform for emotion regulation in parent–child interactions. Pham et al. introduced Galactic Glide, a gamified rehabilitation environment for adolescents undergoing physical therapy. Georgekutty et al. proposed an avatar-based conversational AI for home-based physical rehabilitation, emphasizing long-term engagement and usability. Herrmann contributed a position paper on AI in preventive health from a public health perspective, foregrounding equity, accessibility, and ethical implementation. The workshop concluded with a citizen–researcher panel that enriched the discussion with public perspectives on trust, accessibility, and everyday relevance.

Altogether, the contributions collected in this volume were both promising and inspiring. We look forward to future developments emerging from this work and to continued innovation within the EICS community.

Camille Fayollas
Pieter Van Gorp
Mahmoud Baghdadi
Achim Ebert
Jun Hu
Shah Rukh Humayoun
Sapna Jaidka
Kris Luyten
Tilo Mentler
Philippe Palanque
Parvaneh Parvin
Lucio Davide Spano
Simone Stumpf
Gerrit van der Veer
Luciana Zaina
Jürgen Ziegler

Organization

EICS 2025 Committee

General Chairs

Judy Bowen   University of Waikato, New Zealand
Benjamin Weyers   Trier University, Germany

Technical Program Chairs

Kris Luyten   University of Hasselt, Belgium
Luciana Zaina   Federal University of São Carlos, Brazil

Full Paper and Technical Notes Chairs

Anke Dittmar   University of Rostock, Germany
Célia Martinie   Université de Toulouse, France

Late-Breaking Results Chairs

Sophie Dupuy-Chessa   Université Grenoble Alpes, France
Jessica Turner   University of Waikato, New Zealand

Workshops and Tutorials Chairs

Camille Fayollas   Université Toulouse Capitole, France
Pieter Van Gorp   Eindhoven University of Technology, Netherlands

Doctoral Consortium Chairs

Jürgen Ziegler   University of Duisburg-Essen, Germany
Sapna Jaidka   University of Waikato, New Zealand
Tilo Mentler   Trier University of Applied Sciences, Germany

Demo and Posters Chairs

Nic Vanderschantz   University of Waikato, New Zealand
Max Pascher   TU Dortmund University, Germany

Local Organization Chair

Ivan Blečić   University of Cagliari, Italy

Proceedings Chair

Valentino Artizzu   University of Cagliari, Italy

Web Chairs

Yuen C. Law   Instituto Tecnológico de Costa Rica, Costa Rica
Hongming Zhang   University of Waikato, New Zealand

Publicity Chair

Jemma König   University of Waikato, New Zealand

SIGCHI/ACM Liaisons

Sapna Jaidka   University of Waikato, New Zealand
Judy Bowen   University of Waikato, New Zealand
Benjamin Weyers   Trier University, Germany

Steering Committee Chair

Davide Spano   University of Cagliari, Italy

Steering Committee Members

Alan Dix	Swansea University, UK
Anke Dittmar	University of Rostock, Germany
Benjamin Weyers	Trier University, Germany
Célia Martinie	Université de Toulouse, IRIT, France
Jeffrey Nichols	Apple, USA
José C. Campos	Universidade do Minho, Portugal
Judy Bowen	University of Waikato, New Zealand
Kris Luyten	Hasselt University, Belgium
Philippe Palanque	Université de Toulouse, IRIT, France

EICS 2025 Associated Chairs

Carmen Santoro	CNR-ISTI, Italy
Christian Kray	University of Münster, Germany
Fabio Paternò	CNR-ISTI, Italy
Gustavo Rovelo Ruiz	Hasselt University, Belgium
Jeffrey Nichols	Google, USA
José C. Campos	Universidade do Minho, Portugal
Jürgen Ziegler	University of Duisburg-Essen, Germany
Kris Luyten	Hasselt University, Belgium
Luciana Zaina	Federal University of São Carlos, Brazil
Lucio Davide Spano	University of Cagliari, Italy
Marco Winckler	Université Côte d'Azur, France
Mathias Funk	Eindhoven University of Technology, Netherlands
Peter Forbrig	University of Rostock, Germany
Philippe Palanque	Université de Toulouse, IRIT, France
Radu-Daniel Vatavu	"Ştefan cel Mare" University of Suceava, Romania
Sophie Dupuy-Chessa	Université Grenoble Alpes, France

EICS 2025 - EISEAIT Workshop Chairs

Barbara Rita Barricelli	Università degli Studi di Brescia, Italy
José Creissac Campos	Universidade do Minho, Portugal
Kris Luyten	Hasselt University, Belgium
Sven Mayer	LMU Munich, Germany
Philippe Palanque	Université de Toulouse, IRIT, France
Emanuele Panizzi	University of Rome La Sapienza, Italy

Lucio Davide Spano   University of Cagliari, Italy
Simone Stumpf   University of Glasgow, UK

EICS 2025 - EISEAIT Reviewers

Aaron Conrardy   Luxembourg Institute of Science and Technology,
                                          Luxembourg
Camille Fayollas   Université Toulouse Capitole, IRIT, France
Célia Martinie   Université de Toulouse, IRIT, France
Davy Vanacken   Hasselt University, Belgium
Ezequiel Moreira   Universidade do Minho, Portugal
Gustavo Rovelo Ruiz   Hasselt University, Belgium
Jürgen Ziegler   University of Duisburg-Essen, Germany
Max Pascher   TU Dortmund University, Germany

EICS 2025 - EXDMR Workshop Chairs

Gerrit van der Veer   Radboud University, Netherlands
Achim Ebert   University of Kaiserslautern-Landau, Germany
Shah Rukh   San Francisco State University, USA
Mahmoud Baghdadi   University of Kaiserslautern-Landau, Germany

EICS 2025 - IAI4PH Workshop Chairs

Parvaneh Parvin   Eindhoven University of Technology, Netherlands
Jun Hu   Eindhoven University of Technology, Netherlands

Sponsors

Contents

Engineering Interactive Computing Systems Doctoral Consortium (EICS DC 2025)

An Interdisciplinary Framework for Wearable Air Quality Monitoring Devices

Wensi Cai[(✉)] [iD]

Glasgow Caledonian University, Glasgow, UK
`wensi.cai@gcu.ac.uk`

Abstract. Air pollution is a significant public health concern in many urban environments. A small number of fixed-location monitoring stations provide data that when aggregated and averaged can present a useful picture of an urban environment's pollution levels. However, this is often insufficient for individuals with acute respiratory health conditions. Wearable air quality monitors offer a promising alternative by enabling localized, real-time pollutant sensing. In this research, such monitors are defined as devices that detect airborne pollutants and contextualise these readings against established health standards provided by governmental or international agencies, thereby supporting informed personal decision-making. The development of wearable AQMs faces key challenges in sensor accuracy, wearability, usability, and public trust. This doctoral research addresses these challenges through the development of an interdisciplinary, user-centred design framework that integrates principles from human-computer interaction (HCI), fashion technology, and environmental sensing. The aim is to bridge the gap between technical feasibility and lived user experience, supporting the development of inclusive, acceptable, and practically deployable solutions for everyday air quality monitoring.

Keywords: Human-Computer Interaction · Wearable Devices · Air Quality Monitoring · User-centred Design

1 Introduction

Air pollution is a significant contributor to poor health in many countries, with over seven million premature deaths annually attributed to fine particulate matter (PM2.5/PM10), nitrogen oxides (NOx), and carbon monoxide (CO) [1]. For example, a study [2] in Valencia, Spain, in 2021 showed that 43.7% of mobile sensors exceeded the limit value established by the EU Directive as well as by the World Health Organisation during the assessment period. Many conurbations use a small number of fixed-location AQM systems to provide aggregated and averaged neighbourhood and city-level data. However, this data often lacks the spatial and temporal detail for individuals with acute respiratory health conditions.

One solution is to provide access for these individuals to wearable air quality monitoring (WAQM) devices that provide real-time localized pollution readings in their

C. Fayollas et al. (Eds.): EICS 2025, LNCS 16511, pp. 3–9, 2026.
https://doi.org/10.1007/978-3-032-26051-2_1

immediate vicinity and hence enabling choices about when and how to travel. The design of WAQM devices is a trade-off between multiple factors including the number of pollutants to be monitored, the desired accuracy of the sensors for each pollutant, the range, type and preferred presentation style of the information to be monitored and displayed, the means of communication between sensor, information display and any remote storage device, the energy sources to power the device, the size, shape and fashion appeal of the device, the range of wearable locations for it to be effective, and the comfort in wearing in it [3–6]. Broader adoption also depends on affordability, durability, and whether a device aligns with users' daily habits and style preferences.

The Research Question is:

RQ How can a user-centred design framework be developed to support the creation of wearable air quality monitors that effectively balance sensor accuracy, wearability, and user experience, thereby enhancing adoption in real-world urban environments?

The Research Objectives are:

1. Investigate user needs, environmental contexts, and ergonomic constraints associated with WAQM devices.
2. Analyse existing general purpose, user-centred and participatory design frameworks, identifying limitations in the context of environmental sensing.
3. Design and evaluate an interdisciplinary framework that integrates sensing, wearability, user interface, and aesthetic considerations.
4. Refine and validate the framework through qualitative feedback and prototype engagement, bridging HCI, environmental sensing, and fashion technology perspectives.

The broad research philosophy deployed is pragmatism [7] and then to adopt an inductive reasoning to the research. This approach develops theories based on observations, often using qualitative data and allows for emerging patterns and insights rather than testing predefined hypotheses. Within that approach this study has used mixed methods as its preferred methodological choice. The combining of quantitative and qualitative methods supports the development and refinement of a user-centred design framework.

2 Designing WAQMs

Many people are accustomed to wearing wrist-worn devices such as smartwatches, and may be content to wear a WAQM device, but as a note of caution, wearing more than one device may be an impediment to adoption. Current environmental sensors are portable, but for people on the move often not suitable because they are bulky, uncomfortable and/or look unfashionable. Reviews of wearable devices [8, 9] reveal a gap in understanding about the design factors that influence comfort and everyday usability.

Wearable AQM device platforms [10–12] normally consist of microcontroller units and sensor modules measuring parameters such as temperature, humidity, particulate matter PM2.5/10, carbon monoxide, nitrogen dioxide, and ozone. Commercial products like Atmotube Pro [13] and Flow [14] by Plume Labs illustrate current design trade-offs. While Atmotube Pro supports multiple pollutants, it suffers from bulkiness and limited battery life. Flow prioritises visual appeal and mobile access but has inconsistent sensing performance across conditions. Neither product is tailored to specific user groups such as cyclists, outdoor workers, or people with respiratory conditions.

The practical challenges in designing WAQMs include:

- reduced accuracy in portable sensors that then require regular calibration, especially in environments with changing temperatures and humidity
- shortened battery life from continuous sensing and wireless data transmission that consume significant power
- the size and shape of existing WAQM devices that often make them a poor fit to body shapes and commonly worn clothing
- raw sensor data is often difficult for non-experts to interpret
- the integration of data from wearable sensors with data from fixed and mobile monitoring stations requires sophisticated data processing and analysis to provide a comprehensive understanding of air quality.

Addressing the set of design factors for WAQM devices requires a design framework sufficiently flexible in its coverage to include the integrated concerns of design process, technology design and integration, information management, fashion, and health and safety. Several general-purpose design frameworks, including the Double Diamond [15] and Design Thinking [16], offer valuable iterative processes but often overlook constraints specific to WAQMs—such as battery life, sensor placement, and long-term wearability. Meanwhile, user-centred [17, 18] and participatory models [19, 20] prioritise inclusivity but may lack emphasis on fashion compatibility or hardware modularity.

The focus of this research is to develop a user-centred design framework tailored for WAQM devices, combining principles from user-centred design, participatory design, and fashion technology design to ensure both technical robustness and user acceptability in real-world deployment.

3   Research Method

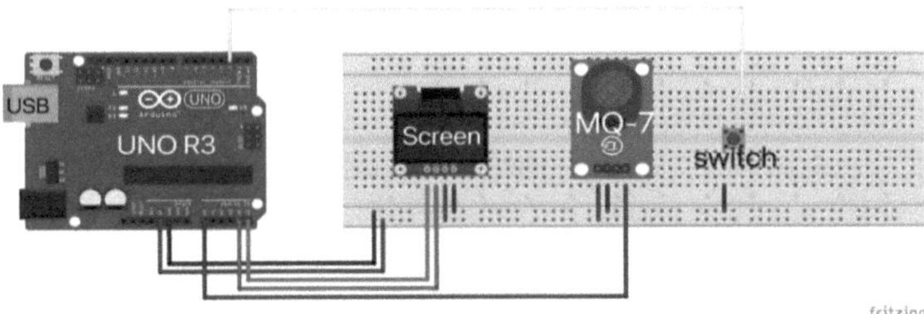

Fig. 1. Prototype using Arduino UNO, CO sensor (MQ-7), display, and switch.

An early-stage prototype was developed (Fig. 1) to explore the construction complexities of a WAQM device such as size, shape, and placement options (e.g., cuff, collar), as well as usability and interaction scenarios. The setup consisted of an Arduino

UNO board, a carbon monoxide (CO) sensor (MQ-7), a small screen display, and a manual switch. The switch activated the sensor, and the screen displayed CO concentration (ppm) in near real-time. While the sensor operated through passive diffusion without active airflow control, it enabled preliminary observations under safe test conditions. The prototype did not aim to deliver calibrated pollutant data but to assess basic technical feasibility and potential integration within wearable formats. Unlike earlier research-grade portable devices—typically hand-held or pocket-carried—this prototype explores design directions for wearable systems intended for continuous, unobtrusive use.

After the prototype was developed, a public-facing questionnaire was made available to gather broad perceptions of WAQM devices, including adoption barriers and feature expectations. The participants were largely based in the city of Glasgow and while not everyone responded with all the requested information, they included cyclists and individuals with respiratory conditions, reflecting a range of typical urban exposure profiles.

In parallel some focus groups were run with design students to discuss reactions to the initial prototype concept and to gather insights on design approaches and preferences to the concept of a WAQM device.

Finally, again in parallel, semi-structured interviews were conducted with seven professionals from environmental science, wearable technology, healthcare, and fashion. These experts do not participate in the earlier public survey and were collectively selected for the range of applied domain expertise they could provide.

4   Initial Results

A full thematic analysis of all data from various sources has not yet been fully completed. This is outstanding work. Table 1 sets out some emergent themes.

Themes were derived through an inductive thematic analysis following Braun and Clarke (2006), involving iterative coding of expert interview transcripts to identify patterns relevant to user needs, technology constraints, and interdisciplinary collaboration

Table 1. Key Themes Derived from Expert Interviews Informing the Design Framework.

Theme	Key Findings
Market Demand	Vulnerable populations (elderly, outdoor workers, cyclists) and tech adopters are primary users.
Technology & Practicality	Trade-offs in sensor size, accuracy, battery life, and user comfort.
Product Development & Testing	Need for iterative real-world durability and wearability testing.
Collaboration & Industry Integration	Importance of multi-sector partnerships (fashion, health, technology).

The market demand may be segmented. Experts highlighted the importance of addressing vulnerable groups such as outdoor workers and older adults. Also, given that there are different technology trade-offs between sensor accuracy, battery life, comfort and real-world durability there can be different cost and price points. Adoption was linked to user-friendly interfaces, aesthetics, and wearability. Some also pointed to opportunities link the data from WAQMs to existing pollution monitoring infrastructure as part of smart city integration. One significant difference in design perspective between this group and the public were the considerations of product engineering i.e. securing the raw materials, manufacturing and assembly complexity, the available labour supply, and the distribution model and cross-sector collaboration particularly across health, fashion, and technology.

Figure 2 shows an emergent conceptual design process framework model that reflects user motivations, expectations, and constraints and recognises the need to balance the technology concerns with fashion concerns.

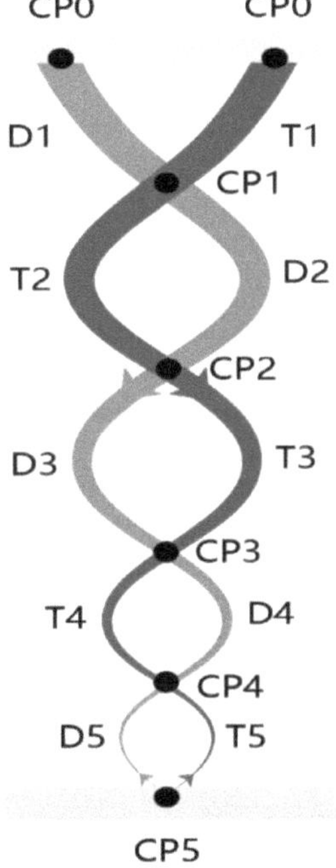

Fig. 2. Interdisciplinary Design Process Model for WAQM Devices, illustrating the technological (T1–T5) and fashion design (D1–D5) development streams. Crossover Points (CP1–CP5) represent integration milestones leading to final product deployment.

The model visualises two intertwined helices—one for the technological stream and one for the fashion design stream—that descend vertically, gradually narrowing toward a final convergence point. Each stream moves through five stages (T1–T5 and D1–D5) and intersects at Crossover Points (CP1-CP5) The final convergence point, CP5, symbolises product launch and market deployment. As each stream moves through the stages, the thickness of the lines reduces progressively, symbolising the reduction of iterative work and consolidation of development maturity. In addition, there are ongoing informal exchanges, reinforcing continuous interdisciplinary collaboration.

5 Summary and Conclusion

This research proposes an interdisciplinary design framework for wearable air quality monitors (WAQMs), grounded in insights from surveys, expert interviews, and prototyping. The framework emphasises three key dimensions: integrating suitable sensors and managing technical constraints such as power and calibration; ensuring wearability through attention to comfort, feedback clarity, and visual design; and supporting co-design processes that involve users, designers, and domain experts. It is modular and iterative, allowing adaptation across different use contexts and design stages.

Next steps include validating the framework through focus groups with UX and fashion design Master's students, whose expertise aligns with usability and aesthetic dimensions of WAQM design.

Disclosure of Interests. The authors have no competing interests to declare that are relevant to the content of this article.

References

1. World Health Organization: Air quality and health. Fact sheet (2021). Retrieved from https://www.who.int/news-room/fact-sheets/detail/ambient-(outdoor)-air-quality-and-health
2. Lorenzo-Sáez, E., Oliver-Villanueva, J.-V., et al.: Assessment of an air quality surveillance network through passive pollution measurement with mobile sensors. In: Environmental Research Letters, IOP Publishing (12 May 2021). https://doi.org/10.1088/1748-9326/abe435
3. Hojaiji, H., Goldstein, O., King, C.E., Sarrafzadeh, M., Jerrett, M.: Design and calibration of a wearable and wireless research-grade air quality monitoring system for real-time data collection. In: 2017 IEEE Global Humanitarian Technology Conference (GHTC), pp. 1–10. IEEE (2017). https://doi.org/10.1109/GHTC.2017.8239308
4. Suhail, A.A., Salman, A.M.: User experience and usability of wearable health devices: a bibliometric analysis (2014–2023). In: 2023 IEEE 13th International Conference on Intelligent Systems (IS), pp. 89–96. IEEE (2023). https://doi.org/10.1109/IS57485.2023.10123456
5. Abayomi-Alli, A., Misra, S., Damasevicius, R., et al.: Sustainability-focused design criteria in wearable health devices. Appl. Sci. 15(1), 461 (2023). https://doi.org/10.3390/app15010461
6. Nibbering, T.: Development of a wearable sensor network for air pollution monitoring. MSc Thesis, University of Twente (2023). Retrieved from https://essay.utwente.nl/100810
7. Dixon, B.: Dewey and Design: a Pragmatist Perspective for Design Research. Springer, Cham (2020). https://doi.org/10.1007/978-3-030-25343-3
8. Masek, A., Zamzuri, N.H., Alias, M.: Design of wearable devices for wearability: a scoping review. Computer. 13(12), 326 (2024). https://doi.org/10.3390/computers13120326
9. Hossain, M.: A scoping review on wearable devices for environmental monitoring. Sustainability. 14(18), 11255 (2022). https://doi.org/10.3390/su141811255
10. Oluwasanya, P., Alzahrani, A., Kumar, V., Samad, A., Occhipinti, L.: Portable multi-sensor air quality monitoring platform for personal exposure studies. IEEE Instrum. Measur. Mag. 22(5), 36–44 (2019). https://doi.org/10.1109/IM-MAG.2019.8804489

11. Geczy, A., Kuglics, L., Jakab, L., Harsanyi, G.: Wearable smart prototype for personal air quality monitoring. In: IEEE 26th International Symposium for Design and Technology in Electronic Packaging (SIITME), pp. 84–88 (2020). https://doi.org/10.1109/SIITME50350.2020.9292309
12. Salva-Pascual, M., Harms, T., Römer, K.: VitalAir: wearable air quality monitoring platform for personal exposure assessment. In: Proceedings of the 19th ACM Conference on Embedded Networked Sensor Systems (SenSys), pp. 125–138. ACM (2022). https://doi.org/10.1145/3560905.3568539
13. Atmotube Inc., Atmotube Pro—Wearable air quality tracker with Bluetooth and app. Product page (2023). Retrieved from https://atmotube.com/products/atmotube-pro
14. Plume Labs, Flow—Personal air quality monitor. Product page (2023). Retrieved from https://plumelabs.com/en/flow/
15. Design Council UK: The Double Diamond. Retrieved April 28, 2025. From https://www.designcouncil.org.uk/our-resources/the-double-diamond/
16. Brown, T.: Design thinking. Harv. Bus. Rev. **86**(6), 84–141 (June 2008)
17. Norman, D.A.: The Design of Everyday Things. Basic Books, New York (2013)
18. Lowdermilk, T.: User-Centred Design: a Developer's Guide to Building User-Friendly Applications. O'Reilly Media, Sebastopol (2013)
19. Chan, J.Y.C., Nagashima, T., Closser, A.H.: Participatory design for cognitive science: examples from the learning sciences and human-computer interaction. Cogn. Sci. **47**(10), e13365 (2023). https://doi.org/10.1111/cogs.13365
20. Rogers, W.A., Kadylak, T., Bayles, M.A.: Maximizing the benefits of participatory design for human-robot interaction research with older adults. Hum. Factors. **64**(3), 441–450 (2022). https://doi.org/10.1177/00187208211037465

Towards Gestural Interaction with Tangible Cubes in Learning Activities

Nacera Latreche[(✉)] [iD]

Université catholique de Louvain, Louvain-la-Neuve 1348, Belgium
nacera.latreche@uclouvain.be
https://www.uclouvain.be/en/people/nacera.latreche

Abstract. Tangible interactions have been widely studied, involving various objects equipped with different sensors in different application domains. Among these, tangible cube interfaces represent a particularly promising modality for these domains, including educational contexts of use, as they promise to improve learning. However, gestural interaction with tangible cubic objects remains mostly task- or domain-specific, limiting their reusability and generalizability. This paper investigates gestural interaction with tangible cubic objects, focusing on AudioCubes, through a mixed-method approach that integrates a systematic literature review, gesture elicitation studies, and usability evaluations. The findings of this study will contribute to the following: (i) a classification of reusable gestures for tangible cubes, (ii) some insights into common barriers for diverse user groups, including children and users with motor or cognitive differences, and (iii) some design implications for more adaptable and generalizable gestural interaction with tangible cubes.

Keywords: Tangible cubes · Gesture input · Learning activities · Tangible user interfaces

1 Introduction

Tangible interactions have generally been widely studied, involving various objects equipped with different sensors in different application domains [5,9, 13,18]. However, despite the diversity of these studies, many existing contributions on gestures for tangible interaction, whether employing a cube or another object, reveal that the gestures are often designed specifically for the tasks of a particular domain [5,18]. For example, consider a student manipulating a cube equipped with sensors in a classroom to visualize data. Tilting the cube to the right, for instance, reveals a new chart. Now, imagine an adult using the same gesture with the cube in a virtual physics simulation to accelerate an object in motion. The physical gesture (tilting the cube) is identical in both cases, yet its interpretation and usability differ across contexts. Gesture patterns may also be

C. Fayollas et al. (Eds.): EICS 2025, LNCS 16511, pp. 10–19, 2026.
https://doi.org/10.1007/978-3-032-26051-2_2

influenced by individual differences, including age, motor skills, cognitive abilities, and personal preferences. For instance, children on the autism spectrum may benefit from structured, predictable, and sensory-appropriate gesture interactions, which may enhance engagement and accessibility. Together, these issues limit the reusability and generalizability of gesture vocabularies for tangible interaction.

Furthermore, less is known today about the use of one particular object, namely a tangible cube, in learning activities [15], thus presenting an opportunity to expand our understanding of such gestures. This expansion would be beneficial for distinguishing between preferred and theoretically possible user gestures. It is necessary to define and follow a method for engineering the tangible interface with this object. This would move away from one-off approaches, which are often valid for a particular case but difficult to reuse or generalize to multiple experiments. Additionally, potential gestures with a tangible cube must be explored as widely and systematically as possible. This broader investigation will contribute to designing gesture sets from a more open and inclusive perspective. In light of these considerations, we investigate the following research question (Fig. 1):

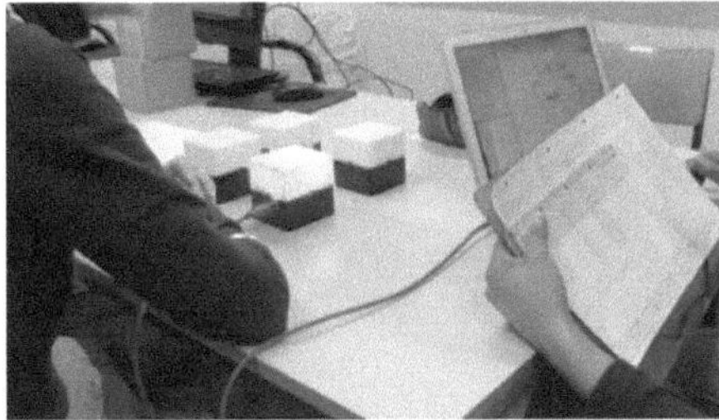

Fig. 1. Two cases of cubic tangible interaction: sound design (left) and learning (right).

RQ = How can cubic tangible interaction be adapted to end-users in learning environments?

The following structure has been adopted for the present doctoral consortium paper. First, we present our research. Then we situate our work within related research and describe our mixed-method approach. Lastly, we discuss our expected contributions.

1.1 Research Question

Following the Goal-Questions-Metrics (GQM) [1], a hierarchical model that follows a top-down approach where a goal is first specified and associated with research questions collected and measured through metrics, framework, we break this question into focused sub-questions, which are aligned with the research program tasks (Table 1):

Table 1. GQM Framework: Understanding and Adapting Cubic Tangible Interfaces

Category	ID	Description
Goal	G_1	Understanding Cubic Tangible Interaction
	G_2	Study user interactions, particularly gesture-based interactions, using the Cubic Tangible Interface (CTI) in general and AudioCubes in different learning contexts.
	G_3	Adapt the Cubic Tangible Interface to end users' needs
Research question	RQ_1	What are the current use cases of CTI?
(**T1**)	RQ_2	How do they differ from other TUIs?
Metrics	M_1	Metrics reported in the PRISMA flow diagram (identification, screening, eligibility, and inclusion)
Research question	RQ_3	What gestures do users naturally perform with AudioCubes in learning tasks?
(**T2+T3**)	RQ_4	How can these gestures be effectively described and classified?
Metrics	M_2	Agreement rate
	M_3	Thinking time
	M_4	Goodness of fit
	M_5	Perceived gesture complexity
Research question	RQ_5	What is the user experience?
(**T4**)		
Metrics	M_6	Task completion rate and time
	M_7	Fatigue
	M_8	Memorability
	M_9	Perceived gesture complexity

2 Related Work

This section reviews (i) tangible interaction with cubes (by application domain), (ii) gesture elicitation studies (GESs) for Tangible Interaction based on our preliminary study.

2.1 Tangible Interaction with Cubes

Kranz et al. developed a cube with six LCD displays and accelerometers for edutainment (picture recognition, vocabulary training, 2D view matching) [11] , emphasizing the interface's hardware setup, gesture-based input, and visual output capabilities showcase the pedagogical potential of cubes but generally do not derive or validate comprehensive gesture sets. Similarly, Zhou et al. introduced Magic Story Cube and Jumanji Singapore [26], two tangible interactive entertainment interfaces using physical cubes combining TUIs with AR for interactive storytelling. They have primarily focused on designing and developing CTI with priority to entertainment over education, while demonstrating how physical cubes could enhance user engagement, and have not systematically explored gesture studies.

Other studies extended cubes into specific educational applications. Camarata et al. employed cubic blocks for spatial navigation in virtual museums [2], allowing users to interact with digital environments in a hands-on manner. In computer science education, Mateu and Alaman [14] used tangible cubes to teach sorting algorithms, transforming complex ideas into concrete, engaging exercises. These studies reinforce the educational value of CTIs and their ability to assist pedagogical activities, but do not explicitly investigate how user gestures should be defined or adapted. Additionally, they do not provide comprehensive usability testing across diverse educational settings.

More recently, He et al. investigated the use of tangible cubes in mixed reality (MR) environments for intuitive data interaction [8]. Their work offers a systematic framework for understanding gesture interactions with CTI, directly addressing the need for targeted gesture elicitation studies in this area. It emphasizes the cubes' suitability for exploring multi-dimensional data due to their maneuverability and stable structure. While the primary focus was on data visualization, the methodologies and insights gained from this work can be applied to educational contexts, particularly in designing intuitive and adaptive gesture-based interactions for learning tools.

In mathematics and physics education, Pioke et al.investigated the use of cubes and block nets in mathematics education [16]. Although no explicit gestures were mentioned, learners engaged in concrete object manipulation. The study reported significant performance improvements, enhanced understanding, and greater accessibility of abstract mathematical concepts, making them more engaging for students. Walsh and Magana [24] explored the use of cubes for learning physics concepts, specifically friction. Learners interacted with the cube through physical manipulation and visuohaptic simulation, though the gestures were not explicitly described or classified. The study demonstrated significant learning gains, particularly in understanding friction variables such as force, speed, acceleration, and distance.

In Music, Sound, and Performance, Schiettecatte and Vanderdonckt [17] explored AudioCubes, which combine spatial manipulation with auditory and visual feedback, enabling gestures like tilting or stacking to trigger specific outputs (e.g., sounds or lights). Their work highlighted the cubes' ability to sense proximity-forest-and-finger-proximity, while promising, this work was confined to sound applications.

2.2 Gesture Elicitation Studies for Tangible Interaction

Valdes et al.conducted a GES asking participants to propose gestures to execute 8 commands (e.g. define and modify a range) for querying big data using Sifteo cubes on a Microsoft PixelSense tabletop [20]. They came up with twelve gestures: cube on surface, click cube, neighbor cubes, pinch in/out, double tap cube on surface, hover cube on surface, circle-tap cube on surface. These gestures were classified according to three properties: space (i.e., in air, on surface, and hybrid), flow(i.e., discrete vs. continuous), and cardinality (e.g., atomic, compound, and parallel). This study is the closest one to our work as it also exploits a GES; however, the conditions differ substantially from ours in terms of device properties and interaction context. In contrast, Wobbrock et al.introduced a methodological framework for gesture elicitation [25]; however, their work focused on surface computing and did not involve tangible objects. Our work builds on these foundations in four key ways: (1)introducing new gesture sets specifically designed for cubic tangible interaction; (2)developing a new classification framework tailored to cubic objects that extends beyond existing taxonomies; (3)Exploring new contexts of use, particularly learning activities involving diverse user groups, including children on the autism spectrum, which have not been addressed in prior GES research; and (4) applying μGlyph notation [4]to formally describe cube gestures, enabling precise, device-agnostic representations that support reproducibility and integration into gesture-recognition systems. This combination provides a more expressive and structured framework for CTI compared to existing elicitation-based or surface-gesture taxonomies. Together, these contributions provide a more comprehensive and reusable approach to gesture design for cubic tangible interfaces.

3   Theoretical, Conceptual, and Empirical Contributions

Piaget's constructivism [10] and Montessori's focus on manipulatives [19] demonstrate that hands-on engagement with physical objects enhances learning, participation, and knowledge retention. This idea is further explained by the belief that learning takes place in a variety of environments, including but not limited to classrooms, lecture halls, auditoriums, and informal spaces such as living rooms, etc. Creating effective tangible interfaces in these different settings requires both technological innovation and pedagogically informed interaction strategies.

To structure this research, we focus on AudioCubes, a distinctive form of cubic tangible user interface developed by Bert Schiettecatte [17]. AudioCubes integrate spatial and auditory modalities, making them well-suited for Human–Computer Interaction (HCI) in learning environments. Unlike conventional cubes, AudioCubes are wireless, battery-powered devices with embedded computing, multi-color light feedback, and sensors that detect proximity, orientation, and interaction with other cubes and objects [3]. Their versatility makes them an ideal artifact for investigating cubic tangible interaction (CTI) in education.

This work aims to consolidate existing knowledge on CTI, develop a conceptual framework, and prepare for empirical validation. To achieve this, we propose a research program organized into four tasks:

Task 1: Understanding CTI:

– Define and conceptualize the interaction space of CTI in general and within the context of learning environments.

- Explore the theoretical foundations and principles underlying CTI, including the group of cubic rotations. The cube group is the one having the largest order of symmetry.
- Investigate how gestures involved in CTI differ from other forms of gestures for tangible interaction, such as those listed in [22].
- Perform a Systematic Literature Review (SLR) on main digital libraries (e.g., ACM, IEEE, ScienceDirect) on references involving a cube as the main tangible user interface. Ensure description of each reference, comparison between them against criteria, and generation of new ideas based on the comparison. Use the PRISMA method.

Task 2: Mastering the AudioCube technology:

- Understand and describe a software architecture for managing interaction with an AudioCube. An AudioCubes can communicate via a MIDI interface. We therefore need to develop a framework for capturing the raw data sent by the AudioCubes via its MIDI interface, to apply gesture recognition algorithms, and to map the recognized gestures to high-level actions, such as those in the taxonomy.
- Develop and exploit MIDI interface and/or an interface to acquire raw data from the device.
- Create new datasets and examples of usage.

Task 3: Fundamentals of Cubic Tangible Interaction: Instantiation to Audio Cubes:

- Perform an initial Gesture Elicitation Study (GES – see [23] for a survey) involving 30 participants for simple Internet-of-Things actions. Identify an initial consensus set of gestures (Fig. 2). Our main target group: children (aged 8 to 12) and children on the autism spectrum in the classroom as the context of use.
- Consider the RepliGES framework [7] to create replications of the initial GES in new configurations [6], such as in different contexts of use, different sets of referents, different sets of participants, perhaps different methods.
- Define and perform additional Gesture Elicitation Studies (GESs) [25] to consolidate the initial one and to determine how many new gestures emerge after each study to create a large-scale evolution model [21].
- Consolidate the results into a classification of CTI gestures. For each entry, give an identifier, a description, and a specification using the μGlyph notation [4]and studying its corresponding representation [12].
- Create a dictionary of cubic gestures with one cube(unimanual dominant or not), two cubes by the same user (bimanual symmetric or bimanual asymmetric), and four cubes hold by two users. Again use the μGlyph notation to define all configurations using the μDEX software (see https://microglyph. imag.fr/)

Fig. 2. Examples of gestures performed with tangible cubes .

Task 4: Experimental Studies of Cubic Tangible Interaction.

- Define and conduct usability testing sessions where participants interact with the AudioCubes, mainly in the area of Learning.
- Collect qualitative and quantitative data on user interactions, feedback, and perceived usability.
- Analyze the data collected to provide a classification of the possible gestures and movements of the audio cube and identify common usability issues, user preferences, and areas for improving the user immersion experience.
- Derive from these results implications for designing and developing CTI, based on AudioCubes.

When performing tasks 3 and 4, quantitative analysis will be conducted using statistical Data analysis methods. A key process metric, thinking time, that represents the delay between the presentation of a task and the start of a movement, will be captured using synchronized video recordings as part of the GES conducted with children in a classroom learning context.

4   Future Contributions

Through mixed-methods research, we will identify preferred gestures and movement patterns, formalize them into a gesture classification framework, and develop a preliminary gesture dictionary. These contributions lay the groundwork for more reusable and generalizable CTI designs, moving beyond task-specific approaches that limit transferability across domains.

4.1 Implications for Engineering Interactive Systems

The gesture dictionary and μGlyph-based notation offer a standardized vocabulary that can be integrated into gesture recognition engines [23,25]. This allows developers to encode gestures using rule-based or machine learning methods. These frameworks facilitate the design of middleware or toolkits that can easily be incorporated into interactive learning platforms, such as Unity, Arduino, and custom Internet of Things (IoT)frameworks, by mapping gestures to high-level educational actions. Additionally, our proposed usability metrics: task completion, fatigue, and memorability-offer engineers practical evaluation protocols to assess system performance across diverse user groups. This approach establishes a clear connection between gesture sets and user adaptability needs, supporting the development of inclusive interfaces that can dynamically adjust interaction complexity to accommodate varying motor and cognitive abilities. This ensures that systems are effective and accessible to a broad range of users. Although this approach was initiated with AudioCubes, yet, it is device-agnostic. This attribute ensures that the engineering insights gained here can be transferred to other cubic tangible systems.

4.2 Example Application Scenario

To illustrate how this approach supports both the definition of gestures and their integration into functional and inclusive educational systems, consider a vocabulary learning activity where learners interact with cubic tangible objects to explore word categories. Using the provided gesture dictionary, an engineer can map a rotation gesture into a navigation between semantic categories (e.g., animals, colors, verbs) and a stacking gesture to confirm a selected answer. As these gestures are formally described with μGlyph notation, they can be directly integrated into a gesture recognizer without reinventing interaction mappings. The usability metrics that we provide, including task completion time and gesture complexity, assist developers in validating whether the implemented system facilitates seamless and accessible interaction for both children and adults, ensuring broad accessibility.

4.3 Next Steps

Future work will refine the gesture dictionary and explore deeper integration of CTIs into interactive learning environments through longitudinal and cross-context studies. By doing so, we aim to contribute to the development of more intuitive, inclusive, and engaging tangible interaction techniques for education.

Acknowledgments. We acknowledge support from the EU EIC Pathfinder Awareness Inside challenge Symbiotik project (1 Oct. 2022-30 Sept. 2026) under Grant no. 101071147.

References

1. Basili, V.R., Weiss, D.M.: A methodology for collecting valid software engineering data. IEEE Trans. Softw. Eng. **10**(6), 728–738 (1984). https://doi.org/10.1109/TSE.1984.5010301
2. Camarata, K., Do, E.Y.L., Johnson, B.R., Gross, M.D.: Navigational blocks: navigating information space with tangible media. In: Proceedings of the 7th International Conference on Intelligent User Interfaces, IUI 2002, pp. 31–38. Association for Computing Machinery, New York (2002). https://doi.org/10.1145/502716.502725
3. Center for Computer Research in Music and Acoustics: Audiocubes project. https://ccrma.stanford.edu/~bschiett/audiocubes/index.html
4. Chaffangeon Caillet, A., Goguey, A., Nigay, L.: µglyph: a microgesture notation. In: Proceedings of the 2023 CHI Conference on Human Factors in Computing Systems, CHI 2023, Association for Computing Machinery, New York (2023). https://doi.org/10.1145/3544548.3580693
5. Fernaeus, Y., Tholander, J., Jonsson, M.: Towards a new set of ideals: consequences of the practice turn in tangible interaction. In: Proceedings of the 2nd International Conference on Tangible and Embedded Interaction, TEI 2008, pp. 223–230. Association for Computing Machinery, New York (2008). https://doi.org/10.1145/1347390.1347441
6. Gheran, B.F., Vatavu, R.D., Vanderdonckt, J.: New insights into user-defined smart ring gestures with implications for gesture elicitation studies, pp. 1–8 (2023). https://doi.org/10.1145/3544549.3585590
7. Gheran, B.F., Villarreal-Narvaez, S., Vatavu, R.D., Vanderdonckt, J.: Repliges and gestory: Visual tools for systematizing and consolidating knowledge on user-defined gestures. In: Proceedings of the 2022 International Conference on Advanced Visual Interfaces, AVI 2022, Association for Computing Machinery, New York (2022). https://doi.org/10.1145/3531073.3531112
8. He, S., et al.: Data cubes in hand: a design space of tangible cubes for visualizing 3d spatio-temporal data in mixed reality. In: Proceedings of the 2024 CHI Conference on Human Factors in Computing Systems, CHI 2024, Association for Computing Machinery, New York (2024). https://doi.org/10.1145/3613904.3642740
9. Ishii, H., Ullmer, B.: Tangible bits: towards seamless interfaces between people, bits and atoms. In: Proceedings of the ACM SIGCHI Conference on Human Factors in Computing Systems, CHI 1997, pp. 234–241. Association for Computing Machinery, New York (1997). https://doi.org/10.1145/258549.258715
10. Kahn, J.H.: The psychology of the child. by jean piaget and bärbel inhelder. london: Routledge and kegan paul. 1969. pp. 173. price 35s. British J. Psychiatry **117**(538), 337–338 (1970). https://doi.org/10.1192/S0007125000193341
11. Kranz, M., Schmidt, D., Holleis, P., Schmidt, A.: A display cube as a tangible user interface. In: Adjunct Proceedings of UbiComp (2005)
12. Lambert, V., Chaffangeon Caillet, A., Goguey, A., Malacria, S., Nigay, L.: Studying the visual representation of microgestures. Proc. ACM Hum.-Comput. Interact. **7**(MHCI) (2023). https://doi.org/10.1145/3604272
13. Lefeuvre, K., Totzauer, S., Storz, M., Kurze, A., Bischof, A., Berger, A.: Bricks, blocks, boxes, cubes, and dice: on the role of cubic shapes for the design of tangible interactive devices. In: Proceedings of the 2018 Designing Interactive Systems Conference, DIS 2018, pp. 485–496. Association for Computing Machinery, New York (2018). https://doi.org/10.1145/3196709.3196768

14. Mateu, J., Alamán, X.: An experience of using virtual worlds and tangible interfaces for teaching computer science. In: Bravo, J., López-de-Ipiña, D., Moya, F. (eds.) UCAmI 2012. LNCS, vol. 7656, pp. 478–485. Springer, Heidelberg (2012). https://doi.org/10.1007/978-3-642-35377-2_66
15. O'Malley, C., Fraser, D.: Literature review in learning with tangible technologies. Tech. rep., NESTA Futurelab (2004). https://telearn.hal.science/hal-00190328
16. Pioke, I., Rivai, S., Talib, S.K.: The use of concrete objects media on students' learning outcomes of cube nets and beams. European J. Humanities Educ. Adv. **2**(12), 93–99 (2021)
17. Schiettecatte, B., Vanderdonckt, J.: Audiocubes: a distributed cube tangible interface based on interaction range for sound design. In: Proceedings of the 2nd International Conference on Tangible and Embedded Interaction, TEI 2008, pp. 3–10. Association for Computing Machinery, New York (2008). https://doi.org/10.1145/1347390.1347394
18. Shaer, O., Hornecker, E.: Tangible user interfaces: Past, present, and future directions. Found. Trends Hum.-Comput. Interact. **3**(1–2), 1–137 (2010). https://doi.org/10.1561/1100000026
19. Tozier, J.: The montessori apparatus: a description of the material and apparatus used in teaching. McClure's Mag. **38**(1), 289–302 (1912)
20. Valdes, C., Eastman, D., Grote, C., Thatte, S., Shaer, O., Mazalek, A., Ullmer, B., Konkel, M.K.: Exploring the design space of gestural interaction with active tokens through user-defined gestures. In: Proceedings of the SIGCHI Conference on Human Factors in Computing Systems. p. 4107–4116. CHI '14, Association for Computing Machinery, New York, NY, USA (2014). https://doi.org/10.1145/2556288.2557373
21. Vanderdonckt, J., Berquin, P.: Towards a very large model-based approach for user interface development. In: Proceedings User Interfaces to Data Intensive Systems. pp. 76–85 (1999). https://doi.org/10.1109/UIDIS.1999.791464
22. Villarreal-Narvaez, S., Sluÿters, A., Vanderdonckt, J., Mbaki Luzayisu, E.: Theoretically-defined vs. user-defined squeeze gestures. Proc. ACM Hum.-Comput. Interact. **6**(ISS) (2022). https://doi.org/10.1145/3567805
23. Villarreal-Narvaez, S., Sluÿters, A., Vanderdonckt, J., Vatavu, R.D.: Brave new ges world: A systematic literature review of gestures and referents in gesture elicitation studies. ACM Comput. Surv. **56**(5) (2024). https://doi.org/10.1145/3636458
24. Walsh, Y., Magana, A.J.: Learning statics through physical manipulative tools and visuohaptic simulations: The effect of visual and haptic feedback. Electronics **12**(7) (2023). https://doi.org/10.3390/electronics12071659
25. Wobbrock, J.O., Morris, M.R., Wilson, A.D.: User-defined gestures for surface computing. In: Proceedings of the CHI Conference on Human Factors in Computing Systems, pp. 1083–1092. ACM (2009). https://doi.org/10.1145/1518701.1518866
26. Zhou, Z., Cheok, A.D., Chan, T., Pan, J.H., Li, Y.: Interactive entertainment systems using tangible cubes. In: Proceedings of the International Conference on Entertainment Computing (2004)

Supporting User Control of User Interface Adaptation with an Extra-UI

Alaa Eddine Anis Sharaoui[(✉)] [iD]

Université catholique de Louvain, LouRIM, Louvain-La-Neuve B-1348, Belgium
`Alaa.Sahraoui@UCLouvain.be`
`https://www.linkedin.com/in/alaa-eddine-sahraoui-018062159/`

Abstract. In context-aware systems, user interface adaptation often occurs automatically, offering limited transparency and minimal user control. To overcome this limitation, this research elaborates on the concept of the Extra-User Interface, an additional user interface layer designed to empower users by enabling real-time observation, customization, and steering of adaptation behaviours. Positioned above the primary user interface, an Extra-User Interface extends the meta-user interface concept by providing structured services, such as inspecting and modifying underlying user interface artefacts. To assess the impact of Extra-User Interfaces on user experience, we adopt a Goal-Questions-Metrics approach, focusing on user satisfaction, comprehension, and performance. The research follows a multi-layered method, combining conceptual development, model-based design, and empirical validation. As a proof of concept, a prototype was developed and tested in a mid-air gesture-controlled environment, showing how an Extra-User Interface can enhance user autonomy and efficiency in controlling adaptation. Moreover, the research explores the applicability of extra-user Interfaces in high-stakes domains through the development of a second prototype for a data visualization interface designed to support critical decision tasks. Contributions of this work are threefold: (1) a theoretical framework defining Extra-User Interface concepts and primitives; (2) a model-based approach for designing graphical user interfaces whose adaptation is controlled by an extra-user interfaces; and (3) the implementation and evaluation of Extra-User Interface environments in realistic settings. Preliminary results validate the feasibility and usefulness of Extra-User Interfaces as a user-centric solution for controllable user interface adaptation in ambient intelligence and ubiquitous computing.

Keywords: Adaptive user interface · context-aware user interface · extra-user interface · gestural interaction · meta-user interface · user control

1 Introduction

Adapting User Interfaces (UIs) to their contexts of use aims at meeting both evolving requirements, such as the needs, desires, and preferences of individual

© The Author(s), under exclusive license to Springer Nature Switzerland AG 2026
C. Fayollas et al. (Eds.): EICS 2025, LNCS 16511, pp. 20–28, 2026.
https://doi.org/10.1007/978-3-032-26051-2_3

users or user groups, and evolving resources, such as the available interaction devices in the surroundings that may all depend on the current situation. There are two categories of adaptation, depending on which entity is responsible for the adaptation: the end-user or the application [10].

Adaptability refers to the end user's ability to adjust the UI, while *adaptivity* or *self-adaptation* denotes the ability of application to perform UI adjustments [2]. *Adaptive User Interfaces* (AUIs) refer to user interfaces benefiting from adaptivity, which can be supported by animated transitions [8]. *Personalisation* or *customisation*, a subset of adaptivity, focuses on adapting the UI contents, presentation, and behaviour based solely on end-user data, such as personal traits [10]. When data comes from external sources, like other user groups, it results in *recommendations* instead. *Mixed-initiative adaptation* occurs when both the end user and the application collaborate to make adaptations [15].

Problem. In all cases, the adaptation process deserves to be under the user's control [9], even in the extreme case where any adaptation can be delegated to the application. To this end, it is necessary to define, design, develop, and test an interface capable of supporting the user's control over this adaptation process, known as an Extra-UI. The aim of this research is to investigate the concept of Extra-UI, to develop a design method for this Extra-UI, and apply and test it. To date, there has been little or no study of this control, and the way to implement this Extra-UI is still in its infancy.

Extra-UIs [7] are explicitly designed as an additional interface layer for **"real-time observation, customization, and steering of adaptive behaviours"**. In other words, an Extra-UI is not just a settings panel for cosmetic changes; it is a dedicated control interface sitting "above" the primary UI that allows users to inspect and modify the *underlying adaptation processes* (e.g. the rules or models that drive interface changes [19]). This concept extends the traditional *meta-user interface* [5] by giving users leverage over the adaptation engine itself, not only the look and feel of the interface, preferably through a dedicated user interface language [14] supporting the preferences [17].

For example, we will clarify that Extra-UIs enable user intervention in the **adaptive logic** such as changing how an algorithm decides to adapt the UI rather than just exposing standard user preferences. This clarification will stress that Extra-UIs support **both** types of adaptation: user-initiated adjustments (adaptability) and system-initiated adaptations (adaptivity), by providing a channel for the user to influence or override automatic behaviours. Recent research on adaptive systems emphasizes that a key design dimension is the **"extent of user control over the adaptation process"** [4], and this work targets increasing that user control and positions Extra-UIs as a unifying layer that facilitates **mixed-initiative adaptation** i.e. a collaboration between user and system in controlling adaptations.

In practice, this means the adaptive system can still make automatic adjustments (adaptivity), but the user has continuous oversight and the ability to intervene or adjust those automatic decisions (introducing adaptability). There is no contradiction because Extra-UIs do not replace automatic adaptation; rather, they augment it with user input, for example the system might normally adapt a layout when the screen orientation changes, but with an Extra-UI the user could set conditions or approve those changes, thereby blending automatic reaction with user preference.

Goal, Research Questions, Metrics. Following the Goal-Questions-Metrics (GQM) [1] framework, a hierarchical model that follows a top-down approach where a goal is first specified and associated with research questions collected and measured through metrics, contributions of this approach are presented in Sect. 2 Goal-Questions-Metrics.

2   Goal-Questions-Metrics (GQM)

To systematically evaluate the impact of the Extra-UI approach on user interface adaptation, we adopt the Goal-Questions-Metrics method. The primary **Goal (G)** is to assess whether providing an Extra-UI improves user control, understanding, and overall user experience in adaptive UIs, without significantly hindering the normal interaction. This goal is refined into specific research questions (Q), each question being associated with measurable metrics (M) (Table 1).

Table 1. GQM Description for Extra-User Interfaces.

Goal (G)		
	Improve user control, understanding, and experience in adaptive UIs through Extra-UI, without hindering normal interaction	
Question (Q)	**Metric (M)**	**Description**
Q1. What is an Extra-UI?	M1.1	Clarity of the definition (expert review)
	M1.2	Completeness of ontological coverage (adaptation operations, strategies, actions)
Q2. How to develop an Extra-UI?	M2.1	Functional coverage of Extra-UI primitives (discover, distribute [13], remodel, parameterize, extend)
	M2.2	Architectural compliance with continuous adaptation loop
	M2.3	Methodological completeness using SPEM (Software Process Engineering Metamodel)
Q3. How to evaluate an Extra-UI?	M3.1	Usability and user experience scores (e.g., SUS, UEQ)
	M3.2	Task success rate and effectiveness of adaptation
	M3.3	User perception of observability and controllability of adaptation

This GQM plan aims to guide the conceptualization, development, and evaluation of Extra-User Interfaces (Extra-UIs). It is structured around three key questions. The first question (Q1) seeks to define what an Extra-UI is, supported by metrics evaluating the clarity of its definition and the completeness of its ontological scope. The second question (Q2) focuses on how to develop an Extra-UI, assessed through metrics that cover service functionality, architectural alignment with continuous adaptation, and methodological completeness using SPEM. The third question (Q3) addresses how to evaluate an Extra-UI, using usability scores, task success rates, and users' perception of adaptation observability and controllability.

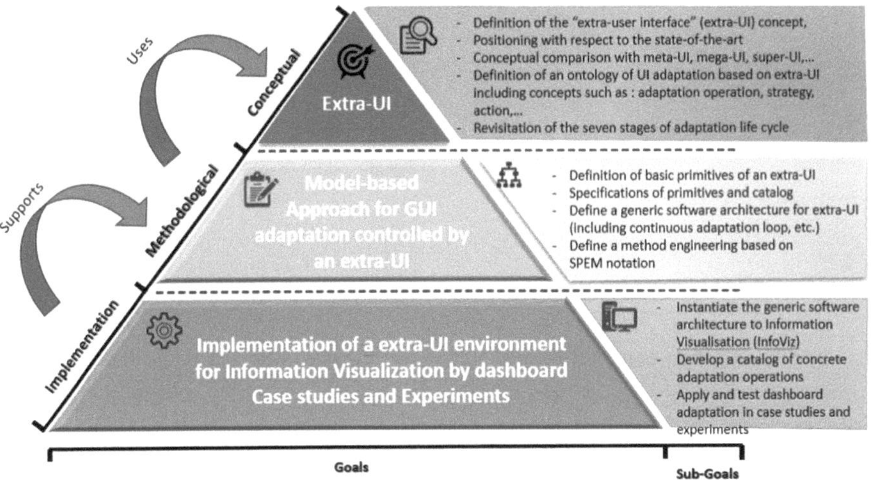

Fig. 1. The three-tiered architecture: progressing from conceptual foundations to method engineering and real-world instantiation.

3   Research Method

This research employs a three-tier approach as represented in Fig. 1:

Conceptual: We define the Extra-UI concept and its theoretical foundation, identifying key services (e.g., discover, distribute, remodel, parameterize, extend) essential for supporting user-controlled adaptation. Extra-UI design guidelines ensure intuitive interaction, with controls presented contextually as overlays or side panels.

Methodological: We develop a model-based engineering method and software architecture, following a structured process for designing adaptive UIs enhanced by Extra-UIs, including clear service integration and adherence to a continuous adaptation loop.

Implementation and Evaluation: Two prototypes are being developed:

1. A **mid-air gesture-controlled environment** using the QuantumLeap framework [18] and Leap Motion Controller. This allows users real-time customization of gesture-command mappings (Fig. 2).
2. A **data visualization dashboard for murder investigations**, enhanced by Extra-UIs enabling investigators to interactively control and fine-tune adaptive visualizations based on machine learning models (Fig. 3). Users can adjust model parameters, visualization thresholds, and interpretability settings directly through the Extra-UI, improving transparency and efficiency.

Empirical evaluation involves comparing interactions with adaptive systems both with and without Extra-UIs, employing usability questionnaires (SUS,

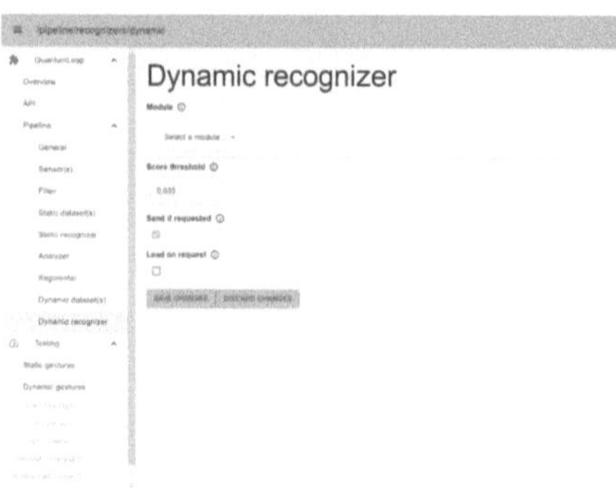

(a) Dynamic recognizer settings.

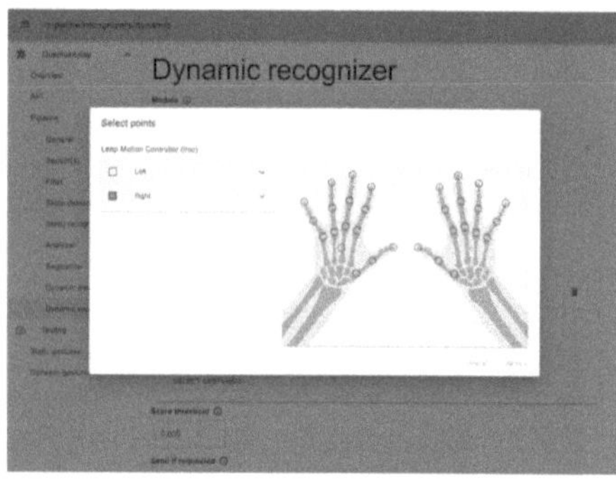

(b) Point selection dialog.

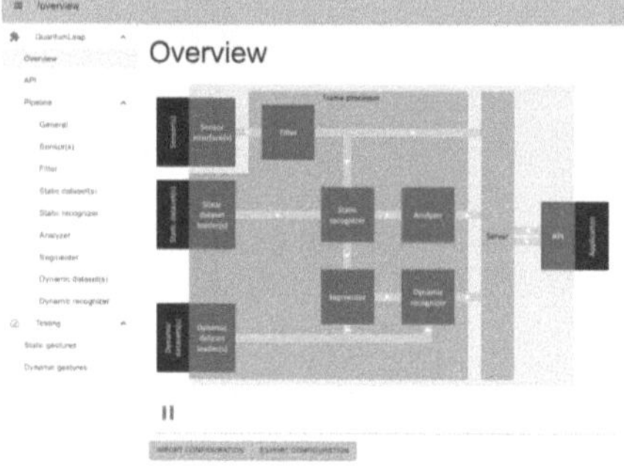

(c) Pipeline overview.

(d) QuantumLeap API (GestureHandler).

Fig. 2. Extra-UI control touchpoints in the QUANTUMLEAP environment: (a) set score thresholds and loading policies; (b) select tracked joints; (c) inspect the adaptation pipeline; (d) programmatic control via API.

UEQ), performance metrics, and user interviews. Preliminary results demonstrate significant improvements in user control, comprehension, and satisfaction without substantial interaction overhead. Iterative feedback continuously informs refinements of the Extra-UI implementations.

4   Current Status and Contributions

The research has several activities in progress:

Conceptual Clarification: We define the Extra-UI concept, situating it relative to meta-UI and mega-UI frameworks, with an explicit focus on enabling runtime user control over UI-level adaptations. This definition helps distinguish Extra-UIs by emphasizing user empowerment and interactive transparency. For example, in a smart home, an Extra-UI might let users adjust lighting automation rules directly [20].

Taxonomy Development: A comprehensive literature review was conducted, resulting in a taxonomy classifying Extra-UIs by the type of objects manipulated (digital vs. physical), integration strategies (embedded vs. external), and provided adaptation services (discover, remodel, parameterize, extend). This taxonomy helps to understand existing solutions and identify research gaps [20]. While prior frameworks such as Meta-UIs and Mega-UIs enable inspection or coordination across interfaces, they rarely provide mechanisms for direct user steering of adaptive logic at runtime. The Extra-UI approach complements these by introducing a controllability dimension across the adaptation loop and by operationalizing user intervention in the decision-making pipeline of the system. This distinction positions Extra-UIs as a human-in-the-loop counterpart to autonomous adaptation, aligning with explainability and transparency.

Prototype Implementation: Two prototypes are developed and evaluated. First, the **Mid-Air Gestural Extra-UI** implementation, based on the QUANTUMLEAP framework, enables users to interactively modify gesture-command mappings in real-time. Initial user studies validated the effectiveness of the prototype, highlighting enhanced usability and adaptability (Fig. 2). Another example is SKETCHIXML [6], which enables users to define how UI elements are recognized, structured and rendered through sketching, gesture training, and grammar editing, thus effectively shaping the UI's adaptation rules. By providing control over input interpretation, rendering fidelity, and widget logic, SKETCHIXML [6] empowers users to influence the interface design process at a higher level, making it a clear example of an Extra-UI focused on low-fidelity prototyping.

This research is also applied on an **Adaptive Data Visualization Dashboard** designed for murder investigations. It integrates Extra-UIs allowing interactive user control over visualization parameters and interpretability settings of machine learning [16] models, specifically Reinforcement Learning (RL) [11,12]. This prototype allows investigators to adjust the settings of a machine learning model or a visualization threshold, which is not analogous to changing a colour theme or font size. Instead, it directly influences the adaptive logic of the system: adjusting a model parameter (e.g. the threshold of a Random Forest for highlighting data points) changes when and how the interface adapts to new data or user actions. In essence, the user is tuning the adaptive behaviour of the system through the Extra-UI. We will make this explicit by rephrasing the example as follows: "For example, investigators use the Extra-UI to fine tune the criteria used in the visualization. The user sends the responses to the adaptive system, instead of configuring the UI appearance." (Fig. 3).

Preliminary Evaluation: Early pilot studies compared interactions with and without Extra-UIs, using usability questionnaires (SUS, UEQ), performance metrics, and user interviews. Results indicate potential improvements in user control, understanding, and overall satisfaction, validating the practicality and potential of the Extra-UI concept. Users reported increasing confidence and perceived transparency when interacting with adaptive systems. Ongoing feedback and iterative refinement processes continue to enhance usability and functionality. Two exploratory studies have been conducted so far. The first involved 12 participants using gestural Extra-UI to remap commands in real time, and the second involved 64 investigators exploring adaptive data visualizations. Both

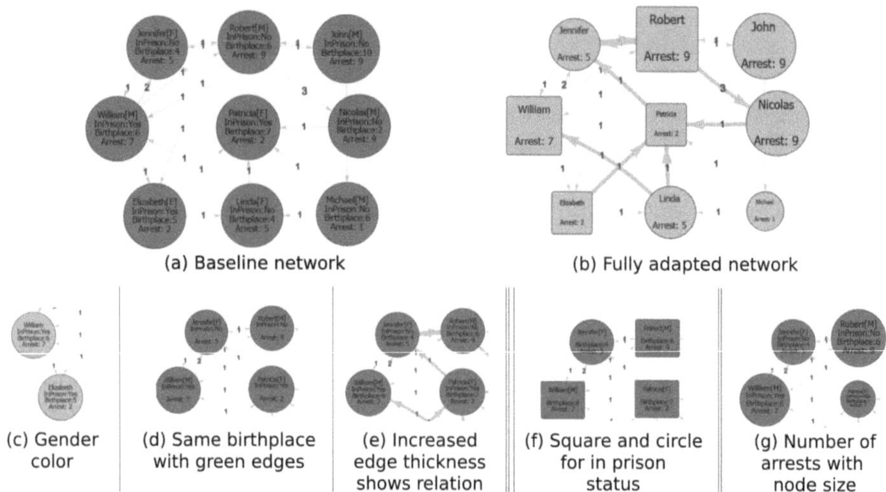

(a) Baseline network

(b) Fully adapted network

(c) Gender color

(d) Same birthplace with green edges

(e) Increased edge thickness shows relation

(f) Square and circle for in prison status

(g) Number of arrests with node size

Fig. 3. Data visualization adaptations with an Extra-UI.

used System Usability Scale (SUS) and User Experience Questionnaire (UEQ) metrics, complemented by interviews on perceived observability and controllability. Future work will expand these studies with larger and more diverse samples to strengthen statistical validity.

5   Conclusion

In conclusion, this doctoral research advocates for and investigates the use of Extra-User Interfaces as a solution to involve end-users in the adaptation loop of interactive systems. By providing an additional interface layer dedicated to observing and controlling UI changes, users are no longer passive recipients of adaptation but active participants who can tailor their experiences. The work so far has laid a strong conceptual and empirical foundation, demonstrating that Extra-UIs can significantly enhance user understanding, agency, and satisfaction in an adaptive mid-air gesture interaction setting. Ongoing and future efforts will broaden and deepen these findings, aiming to establish general guidelines and tools for implementing Extra-UIs across various domains. Ultimately, supporting user control of UI adaptation stands to make intelligent environments more transparent, trustworthy, and aligned with individual user needs – an important step toward truly user-centric adaptive systems.

Acknowledgments. Alaa Sahraoui acknowledges support from UCLouvain and the EU EIC Pathfinder Awareness Inside challenge Symbiotik project (1 Oct. 2022-30 Sept. 2026) under Grant no. 101071147.

References

1. Basili, V., Weiss, D.: A method for collecting valid software engineering data. IEEE Trans. Softw. Eng., **SE-10**(6), 728–738 (1984). https://ieeexplore.ieee.org/document/5010301
2. Browne, D., Totterdell, P., Norman, M.: Adaptive user interfaces. Academic Press, London, UK. https://shop.elsevier.com/books/adaptive-user-interfaces/browne/978-0-12-137755-7
3. Calvary, G., Sahraoui, A.E.A.: Contextual adaptation of user interfaces. In: Vanderdonckt, J., Winckler, M., Palanque, Ph. (eds.) Handbook of Human-Computer Interaction, p. 40. Springer, Berlin (2025). https://doi.org/10.1007/978-3-319-27648-9_17-1
4. Carrera-Rivera, A., Larrinaga, F., Lasa, G., Martinez-Arellano, G., Unamuno, G.: AdaptUI: a framework for the development of adaptive user interfaces in smart product-service systems. User Model. User-Adap. Inter. **34**(5), 1929–1980 (2024). https://doi.org/10.1007/s11257-024-09414-0
5. Coutaz, J. (2007). Meta-user interfaces for ambient spaces: can model-driven-engineering help?. In: Burnett, M.M., Engels, G., Myers, B.A., Rothermel, G (eds.), End-User Software Engineering, 18.02. - 23.02.2007. Dagstuhl Seminar Proceedings 07081, Internationales Begegnungs- und Forschungszentrum für Informatik (IBFI), Schloss Dagstuhl, Germany. http://drops.dagstuhl.de/opus/volltexte/2007/1082
6. Coyette, A., Vanderdonckt, J., Limbourg, Q.: SketchiXML: a design tool for informal user interface rapid prototyping. In: Guelfi, N., Buchs, D. (eds.) RISE 2006. LNCS, vol. 4401, pp. 160–176. Springer, Heidelberg (2007). https://doi.org/10.1007/978-3-540-71876-5_11
7. Demeure, A., Lehmann, G., Petit, M., Calvary, G.: Enhancing interaction with supplementary supportive user interfaces (UIs): meta-UIs, mega-UIs, extra-UIs, supra-UIs. In: Proceedings of the 3rd ACM SIGCHI Symposium on Engineering Interactive Computing Systems (EICS 2011), pp. 333–334. ACM (2011). https://doi.org/10.1145/1996461.1996552
8. Dessart, C.-E., Genaro Motti, V., Vanderdonckt, J.: Showing user interface adaptivity by animated transitions. In: Proceedings of the ACM SIGCHI Symposium on Engineering Interactive Computing Systems (EICS), pp. 95–104 (2011). https://doi.org/10.1145/1996461.199650
9. Eloi, D., Sahraoui, A., Vanderdonckt, J., Leiva, L.A.: User-controlled Form Adaptation by Unsupervised Learning. In: Adjunct Proceedings of the 2024 Nordic Conference on Human-Computer Interaction, NordiCHI 2024, Uppsala, Sweden, 13-16 October 2024. ACM, pp. 16:1-16:8 (2024). https://doi.org/10.1145/3677045.3685431
10. Gajos, K., Chauncey, K.: The influence of personality traits and cognitive load on the use of adaptive user interfaces. In: Proceedings of the 22nd International Conference on Intelligent User Interfaces (IUI), 301–306 (2017). https://doi.org/10.1145/3025171.3025192
11. Gaspar-Figueiredo, D., Abrahão, S., Insfrán, E., Vanderdonckt, J.: Measuring user experience of adaptive user interfaces using EEG: A Replication Study. In: Proceedings of the 27th International Conference on Evaluation and Assessment in Software Engineering, EASE 2023, Oulu, Finland, June 14-16, 2023. ACM. pp. 52-61 (2023). https://doi.org/10.1145/3593434.3593452

12. Gaspar-Figueiredo, D., Fernández-Diego, M., Abrahão, S., Insfrán, E.: A comparative study on reward models for user interface adaptation with reinforcement learning. Empir. Softw. Eng. **30**(3), 109 (2025). https://doi.org/10.1109/ACCESS.2024.3366667
13. Grolaux, D., Vanderdonckt, J., Van Roy, P.: Attach me, detach me, assemble me like you work. In: Costabile, M.F., Paternò, F. (eds.) INTERACT 2005. LNCS, vol. 3585, pp. 198–212. Springer, Heidelberg (2005). https://doi.org/10.1007/11555261_19
14. Helms, J., Schaefer, R., Luyten, K., Vermeulen, J., Abrams, M., Coyette, A., Vanderdonckt, J.: Human-centered engineering of interactive systems with the user interface markup language. In Seffah, A., Vanderdonckt, J., Desmarais, M.C. (eds.) Human-Centered Software Engineering, pp. 129–148. Springer, London (2009). https://doi.org/10.1007/978-1-84800-907-3_7
15. Horvitz, E.: Principles of mixed-initiative user interfaces. In Proceedings of the ACM Conference on Human Factors in Computing Systems (CHI), pp. 159–166 (1999). https://doi.org/10.1145/302979.303030
16. Lopez-Cardona, A., Emami, P., Idesis, S., Duraisamy, S., Leiva, L. A., Arapakis, I. A comparative study of scanpath models in graph-based visualization. In: Proceedings of the 2025 Symposium on Eye Tracking Research and Applications (ETRA 2025) (Article 89, pp. 1–11). Association for Computing Machinery, New York (2025). https://doi.org/10.1145/3715669.3725882
17. Melchior, J., Vanderdonckt, J., Van Roy, P.: A comparative evaluation of user preferences for extra-user interfaces. Int. J. Hum. Comput. Interact. **28**(11), 760–767 (2012). https://doi.org/10.1080/10447318.2012.715544
18. Sluÿters, A., Ousmer, M., Roselli, P., Vanderdonckt, J.: QuantumLeap, a framework for engineering gestural user interfaces based on the leap motion controller. Proc. ACM Hum. Comput. Interact. **6**(EICS), 161:1-161:47 (2022). https://doi.org/10.1145/3532211
19. Vanderdonckt, J., Calvary, G., Coutaz, J., Stanciulescu, A.: Multimodality for plastic user interfaces: models, methods, and principles. In Tzovaras, D. (ed.) Multimodal user interfaces. Springer, Berlin, Heidelberg (2008). https://doi.org/10.1007/978-3-540-78345-9_4
20. Vanderhulst, G., Schreiber, D., Luyten, K., Mühlhäuser, M., Coninx, K.: Edit, inspect and connect your surroundings: a reference framework for meta-UIs. In: Proceedings of the 1st ACM SIGCHI Symposium on Engineering Interactive Computing Systems (EICS), pp. 167–176 (2009). https://doi.org/10.1145/1570433.1570466

Cross-Reality Context Awareness

Sarah Claudia Krings$^{(\boxtimes)}$ (iD)

Paderborn University, Zukunftsmeile 2, Paderborn 33102, Germany
`sarah.krings@sicp.uni-paderborn.de`

Abstract. Recent technological advances enable seamlessly changing between different degrees of reality and virtuality using consumer hardware. This so-called cross-reality brings new chances to solve tasks that augmented and virtual reality alone were not fit for, but it also brings new challenges, for example when users need a mental representation of the content that is currently not visible and when choosing and executing transitions between realities. I propose to use context awareness to lower the mental load in such situations and to improve overall user-friendliness. My thesis begins with a literature research to gain an overview of the specific challenges and existing approaches in cross-reality context awareness. From the results, first, a taxonomy will be developed. There will also be a solution in the form of a framework supporting the inclusion of context awareness into cross-reality experiences. The taxonomy will additionally be the base for cross-reality building blocks that can be used in the framework. The results will be evaluated in developer and end-user studies to ensure that the framework is usable and solves the found challenges.

Keywords: Cross-Reality · Context Awareness · Transitional Reality

1 Introduction and Problem Statement

In recent years, virtual and augmented reality (VR/AR) technology has advanced to a level where it has become possible to transcend the categories of AR or VR and instead seamlessly adjust the degree of reality to the state of the application. Auda et al. [3] call these seamless transitions along the reality-virtuality-continuum (see Milgram et al. [8]) "transitional cross-reality" and define the current degree of reality and virtuality that a user is experiencing as an "Actuality". With all its facets, cross-reality opens up new possibilities and opportunities to improve existing processes. It can be especially helpful for scenarios that are not sufficiently solved by AR or VR alone. An example is space planning, where VR is helpful to gain an overview of an area, while it is necessary to move back into (more) reality to see how planned changes would take effect. Schröder et al. [13] show this in a user study where participants were tasked with placing lighting in a park.

However, (transitional) cross-reality also brings new issues. Since the user can move along the reality-virtuality continuum, there are two different worlds

C. Fayollas et al. (Eds.): EICS 2025, LNCS 16511, pp. 29–38, 2026.
https://doi.org/10.1007/978-3-032-26051-2_4

(the real and the virtual one) that overlap to changing degrees, and it is impossible to fully experience both at once. Users have to remember details that are not visible in their current actuality, leading to an increased mental load. Also, the possibility to seamlessly transition between actualities adds additional navigation tasks on top of the pre-existing ones. Having to ensure that the current actuality is the right one for the planned task parallels the concept of operational modes, which go against user-friendly design [16]. To prevent these issues, it is necessary to increase the overall user-friendliness. Users should not have to manually choose the best actuality for their current tasks and should be supported in keeping a representation of the non-visible actualities without overwhelming them with information.

In the thesis, context awareness with cross-reality-specific representation techniques will be introduced to improve this situation. A framework for including context awareness into cross-reality applications will be developed by first collecting details on issues and possible solutions to them through a literature review. From the resulting taxonomy, different functional building blocks will be prepared, as well as a context awareness engine to bring these blocks together and into the application. This is expected to help lower mental load and increase the user-friendliness of cross-reality applications by mitigating the aforementioned mental load issues. To ensure meeting the goals, user studies will be conducted, also using experiences that were enhanced using the framework.

The following sections will present related work on the topic, followed by the research questions and methodology. Then, the current ideas and results will be described and the current status of implementing these will be described. Finally, there will be a summary of the expected contributions.

2   Related Work

In this section, first, the concepts for cross-reality and context awareness will be defined. Afterwards, the existing related work in the area will be presented in more detail.

2.1   Cross-Reality

Since terminology is often an issue in the area of immersive technologies, this subsection will begin by discussing a fitting definition of cross-reality (CR). Milgram [8] already described a continuum from reality to virtuality, but Simeone et al. [15] coined the term "cross-reality systems" for systems that can seamlessly transition along this continuum. Auda et al. [3] define three types of cross-reality: they call systems transitioning between actualities "transitional cross-reality" and additionally define "multi-user cross-reality" (multiple different users in different actualities) and "substitutional cross-reality" (repurposing an object to a different degree of reality, e.g., typing on a real keyboard in VR). While keeping the other definitions in mind, the focus will mainly stay on transitional cross-reality, since it offers the largest area for applying context-awareness, which will be discussed next.

2.2   Context-Awareness

Context awareness is defined by Dey and Abowd [1] by first defining context as everything characterizing an entity's situation, with an entity being a person, place, or object relevant to the interaction of user and application. There are also other definitions of context, for example, by Schmidt et al. [11], who categorize content into human factors (including the user, the social environment, and the task) and physical factors (including the location, the infrastructure, and the physical conditions). Zimmermann et al. [19] build upon Dey and Abowd's definition by stating context information into five categories: individuality, activity, location, time, and relations. None of the definitions is specific to cross-reality, but they can still be applied to it due to their generality. In the scope of this work, one could add "virtual factors" to Schmidt's context definition to describe what is happening in the virtual environment. Dey and Abowd's definition stays usable, as long as include virtual objects and places are also included. For the following, the term "context-awareness" will be used to describe an application's ability to measure or grasp contexts, in contrast to the user's awareness of their surroundings, which would just be part of the context from this work's perspective. If the context state fulfils certain conditions, parts of the application can be adapted to support the user. The connections between context changes and adaptations can be described as Event-Condition-Action Rules (see e.g. [2]): when an event, such as a context change, is detected, the application evaluates if the conditions for executing an action (adaptation) are met, and, if so, executes said action. There are also other approaches to achieve context awareness, such as the use of artificial intelligence (e.g., see [9]).

2.3   Cross-Reality Context-Awareness

There already are some works applying context awareness aspects to cross-reality, however, the different definitions of cross-reality cause a relatively wide field of solutions between them.

Seo et al.'s [14] GradualReality Interface focuses on improving the interaction with real objects while in VR (substitutional CR) by showing different levels of real-world details, such as displaying a virtual version of a real pitcher's handle on its virtual counterpart when the user reaches for it. Their work also contains transitional CR aspects, as they include a passthrough showing the real world for dexterity tasks like pouring water. They were able to combine VR with real-world objects while still keeping user-friendliness and a high degree of presence.

Wentzel et al. [18] use (transitional) cross-reality to enable VR developers to peek between different realities (namely, a desktop and a VR application) to prevent having to repeatedly remove and put on the headset. The authors focus on the contextual factors of the input device, the viewing device, and the user's memory. For adaptation techniques, they present different so-called "peeking techniques", on one hand for viewing different realities and on the other for sending inputs from one actuality to another.

Li et al. also focus on transitional cross-reality but do so in the context of VR entertainment in a moving car. They examine how adding environmental information in different degrees of reality into the experience influences the awareness that users have of the real world, as well as how their feeling of presence is affected. In their examples, they include texts regarding the current position and/or a live stream of the view outside the VR experience.

Other works also include interesting concepts that could be useful to cross-reality context awareness, but only focus on either cross-reality or context awareness in their work. For example, with their VRception toolkit, Gruenefeld et al. [5] offer a toolkit for using different cross-reality features and conditions while also simulating them in VR. Similarly, Pointecker et al. [10] present and compare different options to transition between AR and VR, and Wang et al. [17] present "RealityLens", a user interface allowing users to peek from VR into the real world, but neither relates them to context awareness. Also, Schmidt and Yigitbas [12] offer a framework for the evaluation of transitional cross-reality environments. On the other hand, context awareness has already been applied to AR and VR separately, for example, Krings et al. [7] presented a development framework for including context awareness in AR applications, providing a framework and several AR-specific pre-implemented context factors.

Overall, while there already are interesting and detailed approaches on both context-awareness and cross-reality separately, only few works combine both and, to the best of my knowledge, there are no comprehensive approaches that focus on offering a range of functionalities. Different definitions of key terms make the results even more diverse. This supports the importance of the thesis goal of offering a comprehensive overview of context-aware cross reality and providing a framework to easily include it into applications.

3   Research Questions

For the reasons mentioned above, a goal of the thesis is to enhance cross-reality by providing well-adjusted context-awareness and representation methods for it. To do so, it has to first be found out (RQ1) "What are the cross-reality-specific challenges for context awareness?". This question shall be answered via the literature research and derived taxonomy. The taxonomy could also help to build a cross-reality specific model or definition for context. Next, using the results from RQ1, (RQ2) "How can the challenges that cross-reality poses for context-awareness be solved?" should be answered. In the process of answering RQ2, special focus will be put on two topics, which seem especially important in the domain of cross-reality context awareness: since cross-reality deals with the issue of having the virtual and the real world overlap, RQ3 asks "How can a representation of the currently non-visible (virtual or real) world be kept?". Also, transitioning between actualities could be confusing and mentally demanding for users. Therefore, individual users should be allowed to adjust their experience as well as possible, leading to (RQ4) "How can end-users be enabled to edit their context-awareness settings?". To check if the solution and implementation are successful in solving the challenges of cross-reality context awareness, they will also be evaluated. Since the framework has to work on both the developer and end-user level, RQ5 "How can the quality of the cross-reality context awareness framework be evaluated?" has to be answered.

4 Methodology

To answer the research questions a methodology was developed for the PhD thesis. It will be described in the following. To gain an overview of the current work in the area of cross-reality context awareness, a systematic literature review will be conducted. Specifically, this literature review will be used to identify the already existing factors of context awareness in cross-reality, from which a cross-reality context-awareness taxonomy will be derived. The study will include any works that contain both "context-awareness" and "cross-reality", or alternative terms for both, and will be conducted across ACM, IEEE, and Springer publications, as well as including results from Google Scholar and Research Gate to ensure nothing is missed. All results will, of course, be checked for validity, as such a broad search spectrum increases the risk of false positives and duplicates. Parallelly, there will be an architecture for a framework for providing context awareness to (cross-reality) experiences containing a general context awareness engine and an editing component providing the actual cross-reality "building blocks" and an Editor to combine them. The building blocks will be derived from the taxonomy created in the literature review. In the next step, the framework and a selection of the building blocks will be implemented for the Unity Editor. In the long term, the framework should be usable for cross-reality independently from specific areas or use cases. The finished framework will then be applied to different scenarios to verify its quality. However, for the early stages of the work and for the remainder of this paper, the focus will be on the first demo case: cross-reality space planning.

The finished framework will also be evaluated using different methods: To get clear, comparative feedback, an A/B developer study will be executed, letting a group of developers use the framework to apply cross-reality context awareness to the space planning application and comparing the results to a group of developers who solve the same task without the support of the framework. Their efficiency (e.g., time to complete their task), effectiveness (e.g., number of added improvements), and satisfaction (using the SUS [4] and TLX [6] Questionnaires) will be measured and there will be interviews conducted. Additionally, the resulting applications could be evaluated with end-users. Regarding end-users, A/B testing in planned as well, giving half of them the cross-reality space-planning application with context-awareness features added using the framework and giving the other half the same application without context-awareness. Again, efficiency (e.g., time to complete their task), effectiveness (e.g., number of mistakes), and satisfaction (SUS [4] and TLX [6]) will be measured to ensure the framework does indeed support users. When the evaluation is finished, the results and all the above steps will be described in detail for the dissertation.

5   Current Ideas and Results

The cross-reality context awareness system will be built with several components, represented in Fig. 1. A rule-based approach was chosen (e.g., in contrast to AI-based options) to keep the framework as transparent and understandable as possible and build upon previous work in the AR domain (see [7]). The *Context-Awareness Engine* is responsible for the functional side of providing context awareness. The details of how context-awareness is applied are settled in the *Context-Awareness Editing* component, which also includes predefined building blocks based on the previously defined taxonomy of cross-reality context-awareness factors. The provided context-awareness is then applied to a cross-reality experience existing in its real and virtual environments. In the *Context-Awareness Engine* (Fig. 1, middle), the *Rules* for context-aware adaptations each consist of a *Condition*, describing a specific set of circumstances, and an *Adaptation* that describes what changes should be executed when the Condition occurs. A Condition has a *Context Factor*, describing what context detail is monitored for change, and a *Value* that the corresponding Context Factor should have for the Condition to apply. Using a composite pattern, multiple Conditions can be stacked into one using logical operators, allowing the expression of more complex states and reuse and combination of existing Conditions. The Adaptations are structured analogously to the Conditions: an Adaptation

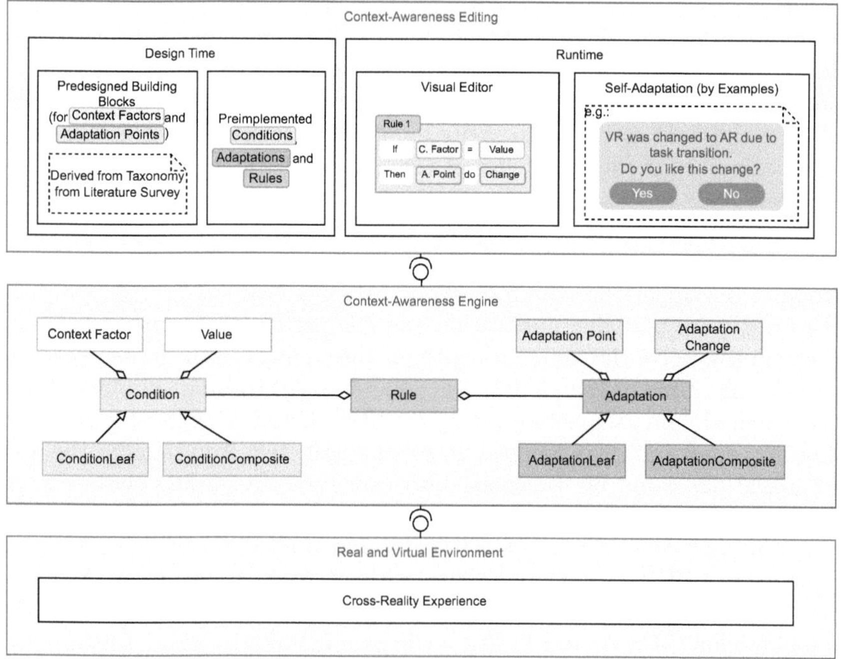

Fig. 1. An Overview of the Planned Context-Awareness System

has an *Adaptation Point* describing the point in the system where the Adaptation should change something and an *Adaptation Change* describing what change should occur on the corresponding point.

To provide context awareness, Rules are needed. Rules are a combination of a Condition (Composite) and an Adaptation (Composite). Textually, they could be expressed as "Rule: IF Condition THEN Adaptation". With Conditions being expressed as "Context Factor = Value" (potentially concatenated with other Conditions using boolean modifiers like AND/OR) and Adaptations being expressed as "Adaptation Point DO Adaptation Change" (also offering boolean operators), the most basic Rule could be "Rule: IF Context Factor = Value THEN Adaptation Point DO Change".

These Rules originate from the Context-Awareness Editing component. The framework offers multiple ways of adjusting the context-aware features: At *Design Time*, there will be *Predesigned Building Blocks* available for Context Factors and Adaptation Points (based on the taxonomy from the systematic literature survey). There will additionally be some pre-implemented Conditions, Adaptations, and Rules provided to lower the entry barrier. At *Runtime*, there are two options for adjusting context-aware features: A low-code *Visual Editor* enabling users to combine the predefined building blocks with values into Conditions, Adaptations, and Rules, and a *Self-Adaptation* option which suggests or context-aware adaptations to the user, enabling them to keep or remove the adaptation rule. The types of editing that are available can be adjusted to the user's skills and mental load using context-awareness.

To illustrate this, the process of integrating this framework into the demonstrator space planning application will be presented: The application allows users to place furniture in the real room when in one actuality, but also offers to use a virtual empty room in a more virtual actuality. The framework could be added to this application to enable different adaptations to improve user friendliness.

For example, when a user places a virtual table in the real room (see Fig. 2), it might collide with some real furniture, making it difficult to imagine what the room looks like when the furniture is replaced with the table. This could be expressed with a condition: the context factor "actuality" is "AR" and the

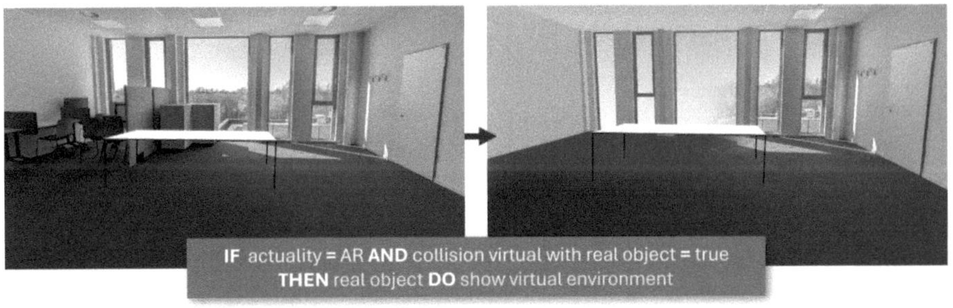

Fig. 2. The Rules and Process of Reacting to a Collision of Virtual and Real Objects

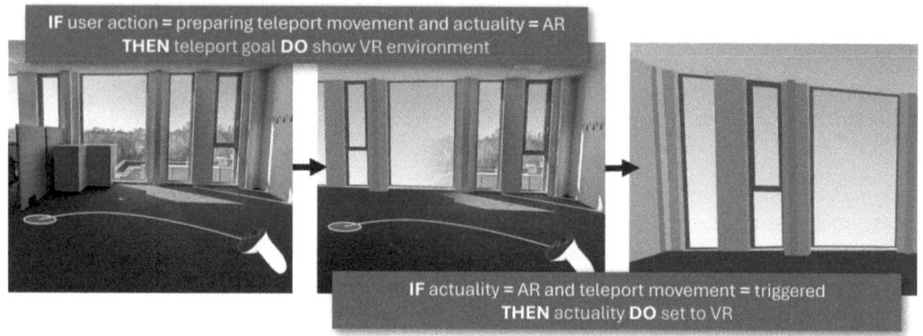

Fig. 3. The Rules and Process for Teleporting from AR

context factor "collision with real object" is "true". Now, a rule could describe an appropriate adaptation to perform in response to this. For example, at the adaptation point "real object", the change to "show the virtual environment" can be activated. This way, the real objects in the collision are covered up by the (empty) virtual room, and the user can see the table without the real objects distracting them, while the rest of the real room is still shown. Without the framework, the user could manually switch to VR or maybe mark an area for VR, which would both require mental effort.

Alternatively, when a user is in the real environment they might want to teleport to a different position (see Fig. 3). This is not possible in AR, so to do this manually, they would have to remember that they can teleport only in the virtual environment, switch to the virtual environment, and then teleport. With the framework, this could be done automatically and even guide the user in two steps to prevent irritating them: the first Rule could be "IF user action = preparing teleport movement AND actuality = AR THEN teleport goal DO show virtual environment", so when a user prepares the teleport action (e.g. by pushing their controller joystick to select a direction), they get a preview of the area of the virtual world that they would teleport into. Then, the second Rule "IF actuality = AR AND teleport movement = triggered THEN actuality DO set to VR", switches to VR fully and executes the teleport when the user selects it (e.g., by letting the controller joystick go to select the position).

6   Status of Doctoral Work

At this point in time, some steps described in the methodology have already been taken. The literature review to determine the cross-reality-specific factors relevant to context-awareness (RQ1) is currently ongoing. The databases and search terms, as well as the data acquisition, are completed, and the review is currently in the full-text screening phase. Parallelly, I have prepared a first draft for the context-awareness framework as described in the previous section. This framework does not only relate to a single research question, but is the basis for the solution of RQ2-RQ4. In addition to the architecture draft, I am also implementing a first draft of a demonstrator in the form of a cross-reality room planning application to test the (also ongoing) implementation of the framework. The finished product will be relevant to research questions RQ2-RQ5, since I also plan to use demonstrators in the evaluation phase. Apart from the evaluation concepts presented earlier, RQ5 currently remains unsolved, since I need a more mature implementation to begin the evaluation process.

7 Expected Contributions

Overall, the thesis will contribute several aspects to the cross-reality domain. The taxonomy of context factors and adaptation points specific to cross-reality combines the current works across literature to give a clear overview of the field and its challenges. It can be used for reference or inspiration as to which factors and actions to consider for a cross-reality application. The second contribution is the framework for enhancing cross-reality applications with context-awareness features to solve challenges such as the issue of representation and transitioning between actualities. Users will additionally be able to control these solutions using the runtime environment. Due to its generic nature, the framework can support the development of most cross-reality applications with fitting context-awareness features, improving user friendliness. Different cross-reality application(s) will also be developed, which, on one hand, demonstrate the framework, but will also offer functionality for their own inherent domains. Lastly, the thesis will offer an extensive evaluation of the proposed solution and give insight and suggestions for the area of cross-reality context-awareness based on the results.

References

1. Abowd, G.D., Dey, A.K., Brown, P.J., Davies, N., Smith, M., Steggles, P.: Towards a better understanding of context and context-awareness. In: Gellersen, H.-W. (ed.) HUC 1999. LNCS, vol. 1707, pp. 304–307. Springer, Heidelberg (1999). https://doi.org/10.1007/3-540-48157-5_29
2. Almeida, E.E., Luntz, J.E., Tilbury, D.M.: Event-condition-action systems for reconfigurable logic control. IEEE Trans. Autom. Sci. Eng. **4**(2), 167–181 (2007). https://doi.org/10.1109/TASE.2006.880857
3. Auda, J., Gruenefeld, U., Faltaous, S., Mayer, S., Schneegass, S.: A scoping survey on cross-reality systems. ACM Comput. Surv. **56**(4) (2023). https://doi.org/10.1145/3616536
4. Brooke, J.: Sus: a quick and dirty usability scale. Usability Eval. Ind. **189** (1995)
5. Gruenefeld, U., Auda, J., Mathis, F., Schneegass, S., Khamis, M., Gugenheimer, J., Mayer, S.: Vrception: rapid prototyping of cross-reality systems in virtual reality. In: Proceedings of the 2022 CHI Conference on Human Factors in Computing Systems, CHI 2022, Association for Computing Machinery, New York (2022). https://doi.org/10.1145/3491102.3501821
6. Hart, S.G., Staveland, L.E.: Development of nasa-tlx (task load index): results of empirical and theoretical research. In: Hancock, P.A., Meshkati, N. (eds.) Human Mental Workload, Advances in Psychology, vol. 52, pp. 139–183. North-Holland (1988). https://doi.org/10.1016/S0166-4115(08)62386-9

7. Krings, S., Yigitbas, E., Jovanovikj, I., Sauer, S., Engels, G.: Development framework for context-aware augmented reality applications. In: Companion Proceedings of the 12th ACM SIGCHI Symposium on Engineering Interactive Computing Systems, EICS 2020 Companion. Association for Computing Machinery, New York (2020). https://doi.org/10.1145/3393672.3398640

8. Milgram, P., Kishino, F.: A taxonomy of mixed reality visual displays. IEICE Trans. Inform. Syst. **E77-D**(12), 1321–1329 (1994)

9. Pichler, M., Bodenhofer, U., Schwinger, W.: Context-awareness and artificial intelligence. ÖGAI J. **23**(1), 4–11 (2004)

10. Pointecker, F., Friedl, J., Schwajda, D., Jetter, H.C., Anthes, C.: Bridging the gap across realities: visual transitions between virtual and augmented reality. In: 2022 IEEE International Symposium on Mixed and Augmented Reality (ISMAR), pp. 827–836 (2022). https://doi.org/10.1016/S0097-8493(99)00120-X

11. Schmidt, A., Beigl, M., Gellersen, H.W.: There is more to context than location. Comput. Graph. **23**(6), 893–901 (1999). https://doi.org/10.1016/S0097-8493(99)00120-X

12. Schmidt, L., Yigitbas, E.: Development and usability evaluation of transitional cross-reality interfaces. Proc. ACM Hum.-Comput. Interact. **8**(EICS) (2024). https://doi.org/10.1145/3664637

13. Schröder, J.H., Schacht, D., Peper, N., Hamurculu, A.M., Jetter, H.C.: Collaborating across realities: Analytical lenses for understanding dyadic collaboration in transitional interfaces. In: Proceedings of the 2023 CHI Conference on Human Factors in Computing Systems, CHI 2023, Association for Computing Machinery, New York (2023). https://doi.org/10.1145/3544548.3580879

14. Seo, H., Yi, J., Balan, R., Lee, Y.: Gradualreality: enhancing physical object interaction in virtual reality via interaction state-aware blending. In: Proceedings of the 37th Annual ACM Symposium on User Interface Software and Technology, UIST 2024, Association for Computing Machinery, New York (2024). https://doi.org/10.1145/3654777.3676463

15. Simeone, A.L., et al.: International workshop on cross-reality (xr) interaction. In: Companion Proceedings of the 2020 Conference on Interactive Surfaces and Spaces, ISS Companion 2020, pp. 111–114. Association for Computing Machinery, New York (2020). https://doi.org/10.1145/3380867.3424551

16. Tesler, L.: A personal history of modeless text editing and cut/copy-paste. Interactions **19**(4), 70–75 (2012). https://doi.org/10.1145/2212877.2212896

17. Wang, C.H., Chen, B.Y., Chan, L.: Realitylens: a user interface for blending customized physical world view into virtual reality. In: Proceedings of the 35th Annual ACM Symposium on User Interface Software and Technology, UIST 2022, Association for Computing Machinery, New York (2022). https://doi.org/10.1145/3526113.3545686

18. Wentzel, J., Anderson, F., Fitzmaurice, G., Grossman, T., Vogel, D.: Switchspace: understanding context-aware peeking between vr and desktop interfaces. In: Proceedings of the 2024 CHI Conference on Human Factors in Computing Systems, CHI 2024. Association for Computing Machinery, New York(2024). https://doi.org/10.1145/3613904.3642358

19. Zimmermann, A., Lorenz, A., Oppermann, R.: An operational definition of context. vol. 4635, pp. 558–571 (2007). https://doi.org/10.1007/978-3-540-74255-5_42

Engineering Interactive Systems Embedding AI Technologies (EISEAIT 2025)

Engineering Large Language Model Agents for Transforming Unstructured Descriptions Into Structured Input

Stefano Zeppieri[1]([envelope]) [iD], Alessandro Aiuti[2] [iD], Alba Bisante[1] [iD],
Venkata Srikanth Varma Datla[1] [iD], Gabriella Trasciatti[1] [iD],
and Emanuele Panizzi[1] [iD]

[1] Department of Computer Science, Sapienza University of Rome, Rome, Italy
{zeppieri,bisante,datla,trasciatti,panizzi}@di.uniroma1.it
[2] Department of Planning, Design Technology of Architecture, Sapienza University of
Rome, Rome, Italy
alessandro.aiuti@uniroma1.it

Abstract. Structured input tasks, such as form filling and data entry, are common across domains but can be time-consuming and cognitively demanding, especially when users must convert free-form descriptions into rigid formats. This paper explores using Large Language Model (LLM) agents to support transforming unstructured natural language and document-based inputs into structured outputs aligned with predefined schemas.

We describe a modular pipeline that takes as input a natural language description, optional supporting documents (e.g., PDFs, spreadsheets), and a form schema. The LLM processes these inputs to generate structured key-value pairs with confidence scores, which can be used to populate forms automatically while allowing users to review and adjust the output. The system follows a mixed-initiative approach, emphasizing human oversight and editable results.

Principles from human-centered AI and adaptive interface engineering guide our design. Rather than presenting a full empirical evaluation, this work contributes a modular architecture and design perspective on embedding LLM agents into structured input workflows, highlighting integration challenges and early feasibility observations in real-world contexts such as the MICS project. The approach is domain-agnostic and compatible with existing infrastructures through the Model-Context Protocol (MCP).

Keywords: LLM · AI · GPT · Usability · UX · Structured Input · Proactive Systems

1 Introduction

The rapid advancement of Large Language Models (LLMs) is reshaping how users interact with digital systems. Beyond their widespread use in content generation, summarization, and retrieval, LLMs offer untapped potential in user interface engineering, particularly in how users input and structure information for downstream processes.

Structured data entry, such as form filling, database population, or system configuration, remains a common but tedious activity across domains ranging from administration to industrial workflows. These tasks often force users to adapt their mental models to rigid input formats, which can be time-consuming and error-prone. Even small inconsistencies in formatting or terminology may disrupt the process, reducing efficiency and increasing frustration. Allowing users to instead express their intent in natural language, and relying on LLMs to translate that intent into structured outputs, can substantially reduce cognitive load and improve usability. Moreover, many workflows involve documents that already contain the required information, such as certificates, datasheets, or resumes. Leveraging these documents through intelligent extraction eliminates redundant re-entry of data and ensures that existing resources are fully exploited.

The promise of LLMs in this space is twofold. First, they can serve as semantic intermediaries, capable of interpreting unstructured user inputs and aligning them with predefined schemas. Second, they enable mixed-initiative interaction, where automation handles straightforward cases while human oversight ensures correctness in ambiguous ones. This balance is crucial: while automation improves speed, user validation preserves trust, accountability, and accuracy. Confidence scores, editable outputs, and transparent reasoning are all mechanisms that support such a human-in-the-loop workflow.

While prior work has explored LLM-assisted form filling, schema alignment, and proactive assistance, many existing systems remain limited to narrow domains or require extensive bespoke engineering to integrate with real infrastructures. A gap remains in demonstrating how LLM agents can be embedded in heterogeneous environments while maintaining transparency and user control.

This work addresses that gap by introducing a domain-agnostic, MCP-compatible architecture that shows how LLMs can operate across diverse datasets and document formats, map them to arbitrary schemas, and support confidence-based mixed-initiative interaction. The novelty lies in the integration strategy, modular engineering design, and emphasis on controllability rather than on proposing a new model.

In summary, our contributions are:

- a modular and domain-agnostic pipeline for LLM-driven structured input generation,
- an integration strategy based on the Model-Context Protocol enabling deployment within existing infrastructures,
- design principles for confidence-based mixed-initiative interaction grounded in HCAI literature.

By building on prior work in implicit interaction, usability evaluation, and context-aware interfaces, we propose a reusable blueprint for embedding LLM-driven structured input generation into interactive systems.

Paper Structure. The remainder of the paper is organized as follows. Section 2 reviews related work on structured input generation, mixed-initiative systems, and human-centered AI. Section 3 introduces our methodological approach. Section 4 describes the engineering of the modular pipeline and the integration of the LLM agent. Section 5 concludes the paper by discussing the contributions, limitations, and future directions.

2   Related Works

Structured input generation is a fundamental yet often labor-intensive task across domains such as administration, healthcare, education, and industrial workflows. Typical activities include form filling, record keeping, and structured data processing. Requiring users to manually input structured data (e.g., by filling out a complex form) rather than allowing them to describe it in natural language is not only tedious but also cognitively demanding. Small mistakes or inconsistencies in formatting can disrupt workflows and introduce errors downstream, underscoring the need for automation and intelligent assistance. Large Language Models (LLMs) offer new opportunities in this space by interpreting natural language descriptions, providing context-aware suggestions, automating completions, and detecting errors, thereby reducing cognitive load while improving accuracy and efficiency [10].

The idea of intelligent assistance for data entry has a long history. Early approaches relied on *rule-based systems*, such as template-driven or keyword-matching auto-fill tools, which were limited to predefined contexts and lacked flexibility. Later, *machine learning (ML)-based systems* introduced personalized predictions and error correction by learning from past user behavior [14]. These methods improved adaptability but still struggled with highly variable or unstructured inputs. More recent work leverages LLMs to interpret free-text inputs and convert them into structured representations. Examples include systems that provide domain-agnostic form filling suggestions [3] and hybrid approaches that combine statistical models with rule-based heuristics for natural language generation and extraction [15]. Compared to traditional ML pipelines, LLMs are particularly well-suited for tasks involving semantic interpretation, schema alignment, and conversational interactions, making them promising candidates for structured input generation.

At the same time, the design of AI-powered interactive systems must adhere to well-established human-AI interaction principles to ensure usability, efficiency, and trust. Amershi et al. [1] emphasize the importance of clarifying AI capabilities, timing assistance appropriately, and maintaining user control and feedback. These considerations are critical in structured input generation, where the risk of over-automation can lead to reduced transparency and user disempowerment. A key design strategy is *mixed-initiative interaction*, as discussed by Horvitz [11], where AI acts proactively when confident but always allows the user to override, refine, or reject its suggestions. In the context of structured inputs, this ensures that automatically generated values remain editable and that users retain control over final data submission.

Beyond interaction design, the Human-Centered AI (HCAI) framework proposed by Shneiderman [13] highlights the necessity of designing AI systems to augment rather than replace human capabilities. Applied to structured input generation, this perspective emphasizes efficiency gains while maintaining transparency and accountability. It also aligns with recent calls in Human-Computer Interaction (HCI) to embed explainability and validation mechanisms directly into AI-assisted workflows, helping users understand and trust the outputs of intelligent systems.

Despite these advances, a research gap remains. Current systems either lack domain generality, rely heavily on predefined rules, or do not provide sufficient transparency for real-world adoption. Few approaches demonstrate how LLMs can be systematically integrated into existing infrastructures to handle natural language descriptions and supporting documents while preserving user oversight. Our work addresses this gap by proposing a modular pipeline that leverages LLMs to transform unstructured inputs into structured data suitable for complex tasks such as form completion, while embedding mechanisms for validation, transparency, and adaptability.

3 Methodology

Building on our studies in need-finding, usability evaluation [4,9], and implicit interaction interfaces [5–8], this research examines how Large Language Models (LLMs) can support structured input generation in real-world workflows. Section 3 presents the methodological foundations of the system. We expand on the design rationale behind the AI agent, the integration strategy using the Model-Context Protocol (MCP), and the mechanisms adopted to improve robustness. We also outline the evaluation dimensions that guide the ongoing assessment of the approach.

Our approach combines LLM-based semantic reasoning with modular integration strategies and user-centered evaluation. In doing so, it enables structured input generation to be applied across domains with minimal reconfiguration, while preserving transparency and ensuring that users remain in control of the final outputs.

3.1 AI Agent Design

The AI agent is designed as a schema-aware extraction component that interprets natural language descriptions and document content and maps them onto the fields defined in the provided schema. The agent combines semantic interpretation, document analysis, and schema alignment in one reasoning process. Figure 1 illustrates how user descriptions and files flow into the agent, which processes them in one pass and produces the structured output that mirrors the schema. The figure is referenced early because it illustrates the conceptual flow guiding all subsequent design decisions.

3.2   Integration with Existing Infrastructures

To support deployment in heterogeneous systems, we rely on the Model-Context Protocol (MCP) [2]. MCP provides a lightweight specification for connecting LLM agents to external tools, files, and application contexts. In our implementation, MCP enables the agent to request form schemas, retrieve documents, and invoke preprocessing modules without embedding system-specific code in the prompt. This separation between model logic and system capabilities makes the approach portable and adaptable to new domains.

3.3   Improving Accuracy and Handling Ambiguity

To improve robustness, we adopt an iterative refinement loop inspired by mixed-initiative principles. The initial extraction is performed using schema-guided prompting. When ambiguities or competing interpretations arise, the agent evaluates multiple candidates and selects the value best aligned with the schema, assigning an appropriate confidence score. Users may edit the output, and their corrections can be reintegrated into later iterations to progressively adjust the prompt configuration and reduce recurring errors. This refinement loop supports transparency and adaptability without requiring complex retraining procedures.

3.4   User-Centered Evaluation

This section outlines the evaluation dimensions guiding the ongoing assessment of the system. While a full user study is part of future work, we adopted estab-

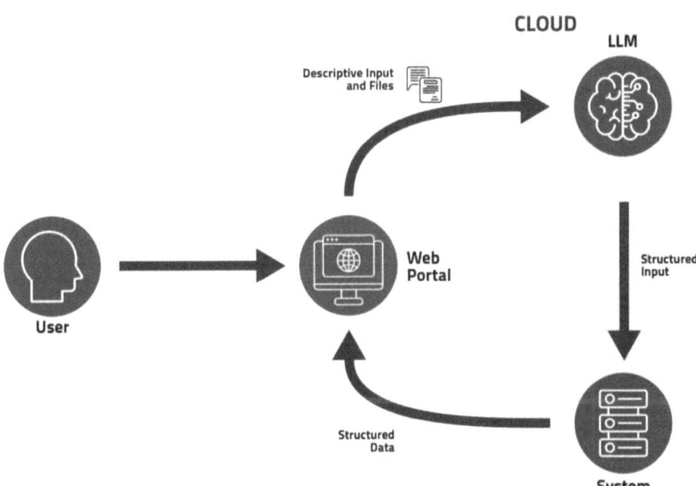

Fig. 1. The LLM receives both natural language descriptions and uploaded documents to generate structured output for form completion.

lished HCI methods for evaluating cognitive load, usability, and trust [1,13]. These dimensions inform the design of the system and structure the feasibility checks conducted within the MICS project.

Evaluation focuses on three dimensions:

- **Cognitive load:** to be assessed through task completion time, error rates, and subjective workload measures.
- **User experience:** to be measured via usability questionnaires and qualitative feedback.
- **Trust and oversight:** analyzed through interactions with confidence scores, editable outputs, and validation mechanisms.

These dimensions guide the design of the system and inform future empirical studies, ensuring alignment with human-centered principles such as transparency, control, and adaptability.

4   Engineering and Modular Pipeline

4.1   Pipeline Overview

The system is implemented as a modular pipeline that connects form extraction, document processing, LLM reasoning, and structured output generation. The architecture prioritizes transparency, portability, and the ability to test each component independently. Figure 2 shows the four main stages:

1. **Form Retrieval:** A script fetches a form's HTML structure given a form ID and a web page URL, and converts it into a structured `form_fields` JSON schema 4.2. This schema defines the fields to be completed, their types, and any validation patterns. From this representation, the system automatically generates a textual description that informs the user about what information is expected and suggests which types of documents (e.g., invoices, resumes, certificates) could support the extraction of the required data.
2. **Input Collection:** The user provides a natural language description of the desired input and may optionally upload relevant documents. To guide this step, the system produces a contextual prompt that helps the user formulate an effective description and select appropriate files. This mixed-initiative approach reduces cognitive load by making expectations explicit and by suggesting what kind of information is most useful.
3. **LLM Processing:** The agent receives as input the `form_fields`, the user's description, and the uploaded documents. These are processed using the `gpt-4o-mini` model equipped with the `File Search` tool. The LLM performs semantic matching between schema fields and the extracted information, resolves ambiguities when multiple candidate values are present, and integrates information across different sources to produce coherent outputs.
4. **Form Completion:** The agent outputs a structured JSON object containing the field `name`, the extracted `value`, and an associated confidence score ranging from 0 to 5. This JSON is then injected into the original form to

complete it automatically. Confidence values guide subsequent user interaction: high-confidence predictions can be accepted directly, while medium- or low-confidence predictions are flagged for review, enabling a transparent human-in-the-loop validation process.

This modular design provides two advantages. First, each stage of the pipeline can be tested and refined independently, facilitating iterative development and targeted evaluation. Second, the architecture supports extensibility, allowing the system to be adapted to different domains simply by changing the schema and prompt configuration rather than the underlying code. By treating the LLM as a context-aware reasoning component and surrounding it with system modules for data collection, interpretation, and integration, we create a cohesive environment where users can interact naturally while benefiting from structured and verifiable outputs.

By structuring the workflow in this way, the system achieves a balance between automation and oversight. It reduces the effort required for repetitive data entry tasks while ensuring that users remain in control of the final outputs. The modularity of the pipeline also makes it portable to different domains, as only the schema and associated prompts need to be adapted to new tasks, without redesigning the entire architecture.

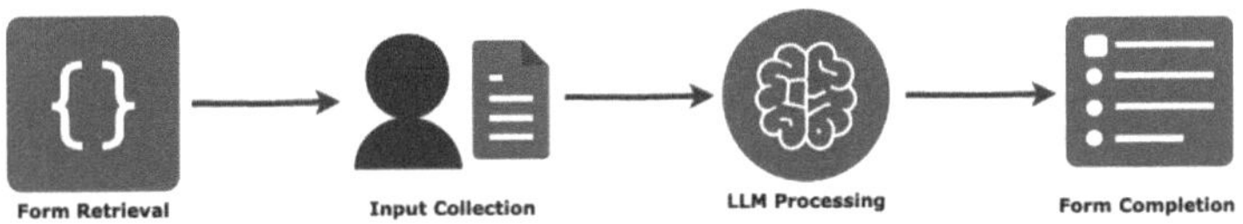

Fig. 2. System Workflow pipeline

4.2   Agent Design and Prompt Logic

At the core of our system lies a specialized LLM agent whose behavior is shaped by a carefully crafted system prompt. This prompt enables the agent to perform three key tasks effectively. First, it interprets the user's intent based on a natural language description, extracting the essential information needed to complete the task. Second, it analyzes the contents of any accompanying documents, such as PDFs or spreadsheets, to identify and extract relevant data that may support or clarify the user's request. Finally, the agent combines these sources to generate a structured output, filling in the appropriate form fields with the most accurate values and assigning a confidence score to each entry.

To ensure reproducibility, we provide a concise overview of the prompt structure used to configure the specialized LLM agent. The full prompt is not included due to its length and implementation-specific formatting, but the core components are stable and can be readily recreated.

The system prompt is structured around three main responsibilities:

1. **Task framing:** defines the agent's role as a schema-alignment assistant responsible for interpreting natural language descriptions and uploaded documents in order to populate the provided `form_fields`.
2. **Schema interpretation:** instructs the model on how to interpret field names, types, validation patterns, and domain hints contained in the `json_schema`. The schema itself defines the expected output structure; the prompt does not specify explicit output formatting rules beyond instructing the model to follow what is defined in the schema.
3. **Confidence scoring logic:** describes how the model should internally assess certainty, how to identify ambiguous or partially supported extractions, and how to assign confidence values from 0 to 5. The schema includes a `confidence` field, which drives the expected format of the model's output.

A short illustrative excerpt is shown below:

"You receive a JSON schema describing the fields of a form, including the structure of the expected output. Use the schema to extract relevant information from the user's description and documents. Populate the fields defined in the schema and assign a confidence score from 0 to 5 based on the reliability of the extracted information."

The expected `json_schema` input format is:

Listing 5.1. JSON Input Schema

```
{"form_fields": [ {"name": "<input name>", "type": "<text
   |number>", "pattern": "<regex>"} ], "description": "<
   natural language description>" }
```

The output is a list of `form_fields` entries with corresponding `value` and `confidence` fields (0–5). The confidence score measures the agent's certainty about each extracted value and serves as a mechanism for guiding user interaction. Values with a confidence score of 5 are considered highly reliable and are used directly to populate the form without requiring user intervention. Scores of 3 or 4 indicate moderate confidence; these values are still used but are explicitly flagged for the user to review and confirm before submission. Entries with scores below three are discarded and not used in the form, prompting the system to either leave the field blank or request additional input from the user.

Given the description: *Hi! I'm Mario Rossi, you can find me at mario.rossi@example.com or call me at +3955511449683.* The agent will return the following structured output:

Listing 5.2. JSON Output Example

```
{"form_fields": [
    {"name": "firstName", "value": "Mario", "confidence":
       5},
    {"name": "lastName", "value": "Rossi", "confidence":
       5},
```

```
    {"name": "email", "value": "mario.rossi@example.com",
        "confidence": 5},
    {"name": "phone", "value": "+3955511449683", "
        confidence": 5}
]}
```

Once generated, the structured output from the AI agent serves as a direct data source for the target system, such as a web form interface, enabling an automatic population of its fields. By aligning the agent's output format with the system's expected input schema, the filled values can be sprogrammatically injected into the form, completing the data entry process with minimal user intervention.

4.3   Implementation and Feasibility Tests

The prototype has been implemented in Python, integrating the `gpt-4o-mini` model with the File Search tool to process unstructured documents. Each stage of the pipeline, form retrieval, input collection, LLM processing, and form, completion was designed as an independent module, allowing isolated testing and refinement.

The early evaluation conducted within the Made in Italy Circolare e Sostenibile (MICS) project [12] serves as a feasibility assessment of the proposed pipeline. The goal of this evaluation is not to measure usability or user experience, but to verify that the system can (i) parse heterogeneous domain documents, (ii) extract schema-relevant values, and (iii) assign confidence scores that correlate with expert judgment. The pilot dataset includes technical datasheets, short textual descriptions, and packaging specifications provided by project partners. For each document set, two domain experts annotated the correct values for each schema field, which served as a reference for assessing extraction correctness.

Preliminary observations indicate that the system can correctly populate a substantial portion of fields in the target schema. High-confidence predictions (≥ 4) frequently aligned with expert annotations, while low-confidence predictions typically corresponded to ambiguous phrasing, incomplete documents, or conflicting sources. Although not intended as a full evaluation, these observations confirm the technical feasibility of the approach and highlight where additional preprocessing and prompt refinement are required. A broader and more controlled evaluation will follow as part of the next phase of the project.

5   Conclusions, Future Work, and Limitations

This paper presented a modular pipeline for transforming unstructured natural language descriptions and heterogeneous documents into structured inputs aligned with predefined schemas. By combining form introspection, document processing, and schema-guided LLM reasoning, the approach demonstrates how intelligent agents can reduce the burden of manual data entry while preserving human oversight through editable outputs and confidence scores.

The preliminary evaluation within the MICS project provided encouraging early evidence of feasibility, particularly in handling domain-specific descriptions and technical documents. These findings suggest that LLM-based structured input generation can support real-world workflows that rely on diverse data sources and rigid schemas.

Several directions remain open for future work. Scaling the evaluation to more diverse datasets and broader project partners will be key to assessing robustness and generalizability. Additional preprocessing-such as OCR for scanned PDFs, table extraction, or domain-specific parsers—could improve handling of noisy inputs. Iterative refinement strategies that incorporate user corrections into subsequent predictions also represent a promising avenue for improving reliability and transparency.

The approach nonetheless has limitations. Confidence scores, while useful for mixed-initiative interaction, are not statistically calibrated and may lead to over- or under-trust. Extraction performance decreases with low-quality or irregular documents, and terminology inconsistencies still introduce ambiguity. Finally, the current evaluation remains preliminary and does not yet include controlled user studies, which will be essential to assess workload reduction, trust, and usability in depth.

Overall, this work provides a practical foundation for integrating LLM-driven structured input generation into existing infrastructures. By outlining both the engineering approach and its current constraints, it aims to support future research on adaptive, transparent, and domain-agnostic systems for intelligent data entry.

References

1. Amershi, S., Weld, D., Vorvoreanu, M., Fourney, A., Nushi, B., Collisson, P., et al.: Guidelines for Human-AI Interaction. In: Proceedings of the 2019 CHI Conference on Human Factors in Computing Systems, pp. 1–13 (2019). https://doi.org/10.1145/3290605.3300233
2. Anthropic: Introducing the model context protocol (2024). https://www.anthropic.com/news/model-context-protocol, Accessed 06 March 2025
3. Aveni, T.J., Fox, A., Hartmann, B.: Omnifill: domain-agnostic form filling suggestions using multi-faceted context (2023). https://arxiv.org/abs/2310.17826
4. Bisante, A., Datla, V., Panizzi, E., Trasciatti, G., Zeppieri, S.: Enhancing interface design with AI: an exploratory study on a ChatGPT-4-based tool for cognitive walkthrough inspired evaluations. In: Proceedings of Advanced Visual Interfaces 2024, AVI 2024. Association for Computing Machinery, New York (2024). https://doi.org/10.1145/3656650.3656676
5. Bisante, A., Datla, V.S.V., Zeppieri, S., Panizzi, E.: Implicit interaction approach for car-related tasks on smartphone applications - a demo. In: Proceedings of the 2022 International Conference on Advanced Visual Interfaces, pp. 1–3 (2022). https://doi.org/10.1145/3531073.3534465

6. Bisante, A., Dix, A., Panizzi, E., Zeppieri, S.: To Err is AI. In: Proceedings of the 15th Biannual Conference of the Italian SIGCHI Chapter, CHItaly 2023, Association for Computing Machinery, New York (2023). https://doi.org/10.1145/3605390.3605414

7. Bisante, A., Dix, A., Panizzi, E., Zeppieri, S.: Implicit interactions in proactive systems: evaluation challenges and adaptations for nielsen's heuristics (2025)

8. Bisante, A., Panizzi, E., Zeppieri, S.: Implicit interaction approach for car-related tasks on smartphone applications. In: Proceedings of the 2022 International Conference on Advanced Visual Interfaces, pp. 1–5 (2022). https://doi.org/10.1145/3531073.3531173

9. Bisante, A., Zeppieri, S., Venkata Srikanth Varma, D., Trasciatti, G., Panizzi, E.: Assessing Large Language Models Adoption in Need Finding: an Exploratory Study (2024), accepted at EISEAIT Workshop, EICS, Cagliari 2024 and soon to be published by Springer in the LNCS series

10. Chiarello, F., Giordano, V., Spada, I., Barandoni, S., Fantoni, G.: Future applications of generative large language models: a data-driven case study on chatgpt. Technovation **133**, 103002 (2024)

11. Horvitz, E.: Principles and challenges of mixed-initiative user interfaces. Commun. ACM **42**(9), 74–82 (1999). https://doi.org/10.1145/330868.330878

12. MICS Consortium: Made in italy circolare e sostenibile (mics). https://www.mics.tech/en/home (2025), Accessed May 2025

13. Shneiderman, B.: Human-centered ai: a new framework for understanding and designing ai systems. In: Human-Centered AI, pp. 1–19. Oxford University Press (2020). https://doi.org/10.1093/oxfordhb/9780190882887.013.3

14. Villena-Román, J., Collada-Pérez, S., Lana-Serrano, S., González, J.C.: Hybrid approach combining machine learning and a rule-based expert system for text categorization. In: The Florida AI Research Society (2011). https://api.semanticscholar.org/CorpusID:15122710

15. Wei, W., Zhou, B., Leontidis, G.: A hybrid natural language generation system integrating rules and deep learning algorithms (2020). https://arxiv.org/abs/2006.09213

Explainable Artificial Intelligence in Critical Interactive Systems: Opportunities and Challenges

Camille Fayollas[1]([✉]) [ID], Moncef Garouani[1] [ID], and Célia Martinie[2] [ID]

[1] IRIT, UMR 5505 CNRS, Université Toulouse Capitole, Toulouse, France
`{camille.fayollas,moncef.garouani}@irit.fr`
[2] IRIT, UMR 5505 CNRS, Université de Toulouse, Toulouse, France
`celia.martinie@irit.fr`

Abstract. Critical interactive systems operate in high-consequence environments where automated decisions have immediate and significant impact. Artificial Intelligence (AI) offers substantial potential to support human operators in these settings, yet its adoption introduces challenges related to transparency, interpretability, and trust. This paper provides a systematic analysis of the opportunities and challenges of embedding Explainable AI (XAI) in such systems, with attention to core properties including usability, dependability, safety, privacy, and confidentiality. We illustrate these considerations through a case study on an AI-driven early-warning system for atmospheric turbulence in commercial aviation, emphasizing the interactions between algorithmic explanations, human cognition, operational constraints, and regulatory requirements. Based on this analysis, we propose research directions that address methodological, human-centered, and regulatory challenges for the practical integration of XAI in safety-critical interactive systems.

Keywords: Explainable AI · Interactive Critical Systems · Task Modeling

1 Introduction

Critical interactive systems (e.g., aircraft cockpits, satellite command and control applications) aim to carry out complex missions. The rise of Artificial Intelligence (AI) technologies may bring new opportunities to support their users in managing their tasks and handling the complexity of these systems. These opportunities will require addressing new challenges to maintain standards in terms of usability, dependability, and safety [20]. In particular, integrating AI in critical interactive systems will require proving that the AI behavior is transparent to the users [14]. Explainable Artificial Intelligence (XAI) has emerged as a crucial field to address the growing need for transparency and interpretability in AI-driven systems [23]. It could thus be a significant means to satisfy the need for transparency in critical interactive systems. However, the integration of XAI will also have to comply with the needs and common practices for the design and development of critical interactive systems [13]. There is thus a need to explicitly identify the opportunities and challenges for integrating XAI in critical interactive systems. In this paper, we present the results of the identification of the opportunities and challenges related to the main properties that the critical interactive system have to fulfill: *usability* to ensure that the interactive parts of the system enable the users to accomplish their tasks, *dependability* to ensure that a failure of the systems or of one of their part will not cause problematic consequences, and safety achieved through the former properties. In addition to these properties, *confidentiality* and *privacy* have also to be taken into account because the explanations provided to the operators are based on data that may not be disclosed to every type of user (e.g., passengers in an aircraft should not have access to the same kind of information as crew members do).

The paper is structured as follows. Section 2 introduces the research background on critical interactive systems and XAI. Section 3 presents the opportunities and challenges for operationalization of XAI in critical interactive systems. Section 4 illustrates these opportunities and challenges with the example of the design of an AI-based atmospheric turbulence early-detection application in a commercial aircraft cockpit. The Sect. 5 presents a selection of research directions and concludes the paper.

2   Research Background

This section first introduces the specific context of critical interactive systems and second introduces the foundational principles of XAI, which together frame the core motivation of this work. Besides our preliminary work [28], we did not find any literature about XAI for critical interactive systems.

2.1   Critical Interactive Systems: Context and Needs

Critical interactive systems refer to interactive systems whose potential cost of system failure is much higher than their design cost (e.g., aircraft cockpits, satellite monitoring and control applications). Engineering these systems combines the specificities from:

- **Engineering interactive systems**: following a user-centered design process [4] to ensure the usability of the system [3].
- **Engineering critical systems**: following a dependable development process to be able to guarantee the dependability of the system (e.g., the development process defined by the DO-178C standard [5] for avionics systems). This requires identifying and analyzing threats and faults that may impair systems' functioning. It also requires developing and integrating several types of mechanisms to prevent, remove, tolerate, or forecast these faults [10]. To be operated, these systems need to be certified. This is supported by standards for ensuring a certain level of safety for systems, software, and their

deployment in an operational context. For example, the Certification Specification 25 for large aeroplanes [1] defines the requirements that are taken into account for large aeroplanes' airworthiness in the European sky. This standard states, in its section 1302, that "...installed equipment must be shown...to be designed so that qualified flight-crew members trained in its use can safely perform their tasks associated with its intended function...". This statement highlights that the interactive system functions have to match user tasks and that users are trained to use the system. The standard also specifies, in its section 1309, the requirements for avionics systems' dependability, defining the probability of failure of a system, depending on its criticality (for instance, a catastrophic failure must be extremely improbable, which means a quantitative probability of failure condition inferior to 10e−9 per flight hour).

2.2 Explainable Artificial Intelligence

XAI is an active research area that focuses on making the reasoning processes of "black-box" AI systems transparent and understandable to human users [23]. As AI models become more complex, particularly with the widespread adoption of deep learning, reinforcement learning, and ensemble methods, the need to ensure that these models remain interpretable to developers, regulators, and end-users has become paramount [21]. XAI techniques can be broadly divided into two categories [18]:

- **Intrinsically interpretable models**, such as decision trees or linear models, which are understandable by design but may lack expressiveness for complex tasks.
- **Post-hoc methods**, which aim to explain opaque models like neural networks after predictions are made. Examples include feature attribution methods (e.g., SHAP, LIME), counterfactual explanations (e.g., "what-if" scenarios), and example-based reasoning.

The literature on XAI has proposed several types of explanations, each serving different cognitive and operational purposes depending on the user's goals, the task context, and the temporal constraints [22]. The most commonly identified types include:

- **Why explanations**: Justify why a particular decision or prediction was made by the system. These are fundamental for post-hoc rationalization and retrospective analysis.
- **Why-not explanations**: Highlight why alternative outcomes were not chosen, which is crucial for understanding boundary conditions or model preferences.
- **What-if explanations & counterfactuals**: Explore how changes in input features would alter the outcome. These explanations are useful for sensitivity analysis and scenario simulation.
- **How explanations**: Describe the internal logic or decision path of the system, especially in rule-based or symbolic models.

Despite the progress made, most XAI methods are developed in static, offline contexts with limited consideration of real-time interaction, domain-specific constraints, or human cognitive load [23].

3 Operationalizing XAI in Critical Interactive Systems

The integration of XAI into critical interactive systems introduces both significant opportunities and complex challenges that are presented in the next subsections.

3.1 Opportunities of Operationalizing XAI in Critical Interactive Systems

XAI is a cornerstone for embedding AI technologies in critical interactive systems [17]. Beyond mere transparency, operationalizing XAI means turning explanations into actionable components that support engineering rigor, user trust, and system certification. Good explanations can thus enable several key opportunities across different layers of the system lifecycle, these opportunities are detailed in the next subsections.

On the Engineering Side. Explainability offers a powerful tool for understanding and managing model behavior throughout the system's development and deployment. It allows engineers to identify the model's functional boundaries, uncover causal relationships between inputs and outputs, and detect undesirable biases that may compromise fairness or robustness. Explanations also help characterize model confidence and uncertainty, which are critical to designing safe fallback mechanisms, validating performance under distributional shifts, and guiding iterative model refinement. In addition, explanation artifacts can be integrated into system documentation, supporting traceability and continuous verification throughout the AI development lifecycle.

On the End-User Side. Explainability is essential for fostering appropriate levels of trust, enabling human oversight, and supporting effective human–AI collaboration. Well-designed explanations empower users to understand why specific decisions were made, to detect and report potential errors, and to calibrate their reliance on automated recommendations. In interactive or safety-critical contexts—such as medical decision support or air traffic control—explanations can reduce cognitive load and facilitate situational awareness by linking system behavior to user goals and domain knowledge. Ultimately, explainability helps bridge the gap between algorithmic reasoning and human understanding, turning AI into a transparent and dependable teammate rather than a black-box assistant.

On the Certification Side. Explainability opens new perspectives for system assurance and regulatory compliance, which are particularly stringent in safety-critical domains.

- *To support compliance with safety requirements*: explanations can strengthen confidence for certification authorities by making AI decision pathways auditable and interpretable. They can also help operators understand the limits of automation, identify safe operating conditions, and anticipate potential failure modes.
- *To support compliance with legal requirements*: XAI contributes to meeting transparency and accountability obligations, such as those defined in the General Data Protection Regulation (GDPR) [2]. Beyond ensuring that certain sensitive attributes (e.g., race, gender, or health data) are not used directly in decision-making, XAI can help detect whether models are inadvertently relying on proxy variables correlated with protected attributes, which could lead to indirect discrimination [36]. By providing intelligible explanations of model decisions, XAI enables auditing and verification that outcomes affecting individuals are fair and compliant with legal requirements [31]. This capability goes beyond the baseline GDPR restrictions by offering practical tools to *inspect, monitor, and justify* algorithmic behavior in operational settings [21].

Beyond these distinct opportunities, operationalizing XAI in critical interactive systems promotes a systemic view of explainability—as a property that cuts across engineering, interaction, and regulation. It encourages the design of AI systems whose explanations are not add-ons but integral components of dependable and trustworthy system behavior. Such an approach paves the way for context-aware, uncertainty-sensitive, and certifiable AI systems capable of operating safely and transparently in high-stakes environments.

3.2 Challenges for Operationalizing XAI in Critical Interactive Systems

Despite its promises, embedding XAI in critical interactive systems presents significant challenges due to both contextual constraints and intrinsic limitations of current explainability methods. These can be linked to the two main properties of critical interactive systems that have been presented in Sect. 2 (usability and dependability), but also to security issues (confidentiality and privacy). These three categories are detailed in the following subsections.

Challenges Linked to Usability. We identified four main challenges linked to usability that are detailed in the next subsections: i) providing explanations relevant to user tasks and user type, ii) providing explanations relevant to temporal constraints, iii) engineering of interfaces and interactions, and iv) modifying users' training.

Providing Explanations Relevant to User Tasks and User Type. Current XAI methods often focus on mathematical fidelity or local feature attribution, but fail to produce meaningful explanations that support user tasks in complex, high-risk scenarios. We need to develop techniques and tools to support the elicitation of explanations that will be the most relevant to mission tasks and the possible contexts in which they could be provided. Moreover, different types of explanations may be perceived differently by users. [24] highlights that some types of explanations help users decide more quickly and that other types can lead to higher levels of perceived transparency and overall satisfaction. Figure 1 presents the relationships between the properties of explanation types that emerged from the experimental comparison of ten types of explanations [24].

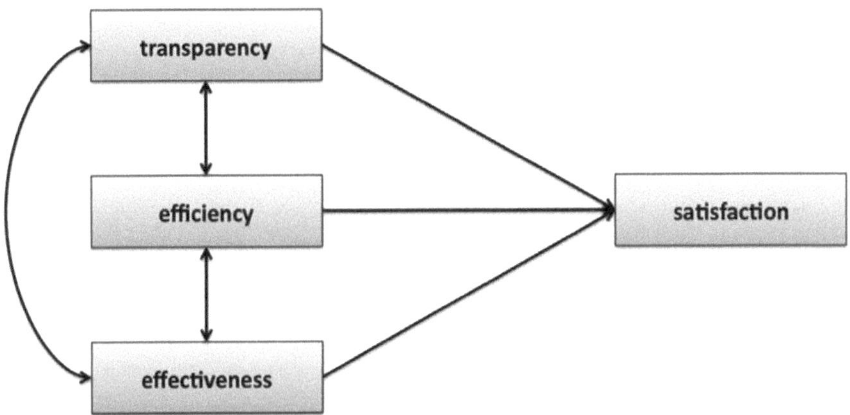

Fig. 1. Relationships between the properties of explanation types (reprinted from [24], with permission from Elsevier)

It is important to note that the main properties of the usability standard (effectiveness, efficiency, and satisfaction) [3], which highly depend on user tasks, are key for assessing the types of explanations. Moreover, [24] also highlights that transparency and persuasiveness are two additional key properties. Transparency is the capability of a system to expose the reasoning behind it to its users [26]. Persuasiveness is the ability of an explanation type to convince the user to accept or disregard certain items [12]. To reach these two additional properties also requires focusing on user tasks, particularly by ensuring that the explanation provides specific content data (closely related to the task), as well as uses information the user is already familiar with [24]. About the user profiles, although some studies indicate that user characteristics do not have a significant effect on the understanding of explanations [30], other studies, such as [15], show that recommendations should adapt to the users' mental model. We believe that relevant user characteristics (e.g., user's existing experience and skills, prior beliefs and potential biases) are important, but the user role has a higher impact. The three main types of roles that have been identified are [32]: developers, domain experts, and end users. However, this classification is quite abstract, and roles should be refined when adapting AI explanations [16].

Providing Explanations Relevant to Temporal Constraints. Temporal constraints play a crucial role in real-time critical systems [9]. In situations where users must react immediately such as handling unexpected events in a vehicle cockpit, the explanation must support rapid understanding of the environment and potential action consequences within strict time limits. A key unresolved issue concerns the *computational latency* of predictions and explanations. While modern models may deliver inference in milliseconds, explanation generation especially for perturbation-based methods like SHAP or LIME can require hundreds or thousands of model queries, making them unsuitable in strict real-time contexts. Even model-intrinsic methods (e.g., gradient- or attention-based explanations) generally introduce non-negligible overhead. Thus, even when prediction time is acceptable, the combined latency of prediction, explanation, and human interpretation may exceed operational deadlines [37].

Two aspects must therefore be considered [9]: selecting explanation types compatible with temporal bounds, and supporting users in maintaining temporal awareness during explanation analysis. Overall, deploying XAI in critical systems requires explanation methods with predictable, bounded execution times, and interfaces enabling users to extract just-enough information under time pressure.

Engineering of Interfaces and Interactions. Existing XAI interfaces offer limited support for interactive exploration and tailored information access [22], and limited support to ensure scalability of explanations in case of complex systems. The need for specific interfaces and interactions is well illustrated by the guidelines for explanations defined by [15]. These guidelines, summarized in Fig. 2, are divided in four principles:

- Natural language: explanations should be written in natural language, as visual explanations can only be understood by experts.
- Responsiveness: explanations should enable the user to respond, for example, with follow-up questions. This is illustrated by progressive disclosure, allowing follow-up on initial explanations.
- Different explanation methods: offering multiple explanation methods and modalities helps understanding in many ways.
- Adaptation to mental models: explanations should adjust to the user's mental models and contexts.

While the last principle is directly related to our first point (providing explanations relevant to user tasks and user type), the first three illustrate the need for specific interfaces and interactions:

- Natural language and different explanation methods. User interfaces and interactions are needed to enable the display of the explanations in different ways. For instance, [38] presented a switch allowing the user to convert a visual explanation into a detailed text.

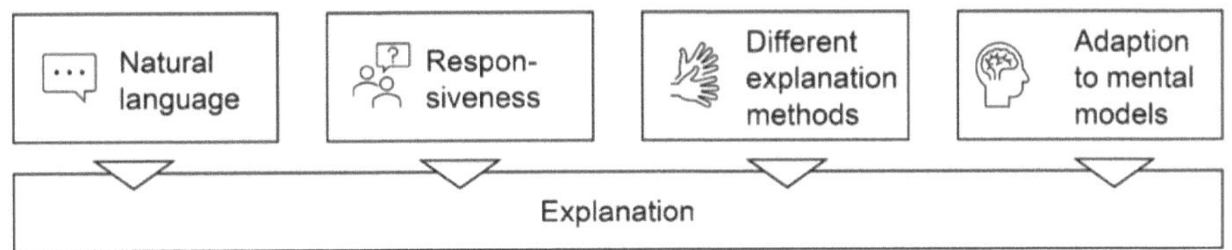

Fig. 2. Design principles for explanations from [25]

– Responsiveness through progressive disclosure. [34] recommends using progressive disclosure by offering hierarchical or iterative explanations following up on initial explanations. This calls for specific interfaces and interactions, such as the "why?" button presented by [29] that enables the display of a one-dimensional visual explanation that can be further detailed in a multi-dimensional visual explanation when clicking on another button.

Modifying Users' Training. Users of critical interactive systems learn to apply specific procedures according to specific contexts, in order to increase operators' performance and to decrease the number of potential human errors when using the system [6]. Training programs aim to provide operators with a predefined set of skills and knowledge before using the system. Embedding XAI in critical interactive systems will require identifying which user tasks are concerned with understanding AI functions' output and their explanations, as well as identifying the most appropriate means to teach XAI analysis to the users in the context of their mission. The training program will thus require implementing these means to train the users to be able to recognize, understand explanations, and to use them to achieve their goals.

Challenges Linked to Dependability. We identified two main challenges linked to dependability that are detailed in the next subsections: i) reliability, and ii) providing evidence during the certification process.

Reliability. XAI components' probability of failure must comply with the corresponding function's criticality. For instance, [7] shows that, in the case of a railway signal reading system, the current human-operated system leads to an average target of a maximum of $10e-5$ failures per signal. Therefore, an AI-based system should achieve the same average of failures per signal. However, the best automatic reading system achieves on average $10e-2$ errors per signal.

XAI functions may help define more reliable AI algorithms. For instance, [8] shows that the use of XAI techniques such as SHAP helps deliver a Data-Driven Turbulence Model with exceptional accuracy. However, the current reliability of XAI functions is constrained by several factors, such as the inadequate handling of uncertainty and ambiguity, the level of confidence in the algorithms' output, and the sensitivity to input data [33].

Providing Evidence During the Certification Process. For certification purposes, providing explanations can help with two matters:

- Give confidence to the authority. Understanding the system enables assessing it with a correct understanding of its functioning and its motivations [17]. However, this implies that the given explanations are precise, clear, and correct with respect to the target domain, the safety of corresponding users' operations, and the certification authorities' tasks and goals. This also implies finding ways to present these explanations to the certification authorities and to prove the reliability of XAI functions.
- Support compliance with legal requirements. XAI is also required for the sake of ethical reasons [17], such as the "right to explanations" regarding automated data processing from the General Data Protection Regulation (GDPR) [2]. This implies being able to define precise, clear, and correct explanations to both GDPR authorities and end-users.

To conclude, to use XAI functions to help provide evidence during the certification process, current challenges rely on both proving the correctness of the XAI algorithms and delivering clear and concise explanations with respect to the end-user's tasks and goals (the end-user corresponding here to certification authorities), which relates to our first identified challenge.

Challenges Linked to Confidentiality and Privacy. In critical interactive systems, explanations are not neutral objects: they may reveal sensitive information about the underlying predictive models, their training data, or even operational parameters. Providing transparency to support human operators must therefore be carefully balanced with preserving the confidentiality of data sources (e.g., operational records, passenger information, or proprietary industrial datasets) and respecting privacy regulations [19]. Moreover, different categories of users (e.g., operators, engineers, regulators) may have different rights to access and interpret explanations, raising the question of access control for explanation content. XAI methods should thus be designed to deliver task-relevant insights without disclosing information that could compromise safety, security, or data protection. This requires new approaches for explanation filtering, abstraction, and role-based tailoring to ensure that explanations remain both trustworthy and compliant with confidentiality and privacy requirements [35].

4   Illustrative Example

To ground the discussion, we consider the case of turbulence modeling in aeronautical systems, as presented in [8]. This example illustrates both the opportunities and challenges of operationalizing XAI in critical interactive systems.

4.1   Case Study Description

[8] developed a data-driven neural network-based turbulence model in several flow configurations where conventional approaches perform poorly. The proposed neural network-based turbulence model predicts the turbulence fluctuations using the Reynolds stress and the mean flow quantities. Furthermore, the neural network model is enriched with SHAPley Additive Explanations (SHAP) to complement the predictive power of the neural network with explainability. The neural network model is based on the FFNN architecture and is developed to minimize the error between the true output and the model output. The model's goal is to find the optimal weights and biases for a given input feature. As shown in Fig. 3, the model takes as input four features (mean velocity gradient tensor, turbulent kinetic energy gradient, mean pressure gradient, and Reynolds number) and computes the Reynolds stress. In addition to this study, the researchers studied the effect of input feature selection on the model's accuracy and conducted an importance analysis using SHAP, which corroborated the findings regarding the effect of input features in their study [8].

This study is of major interest as severe atmospheric turbulence is the leading cause of injuries in civil air transports [27], and real-time identification of atmospheric turbulence as proposed by [27] may enable alerting flight crew and passengers of future turbulence.

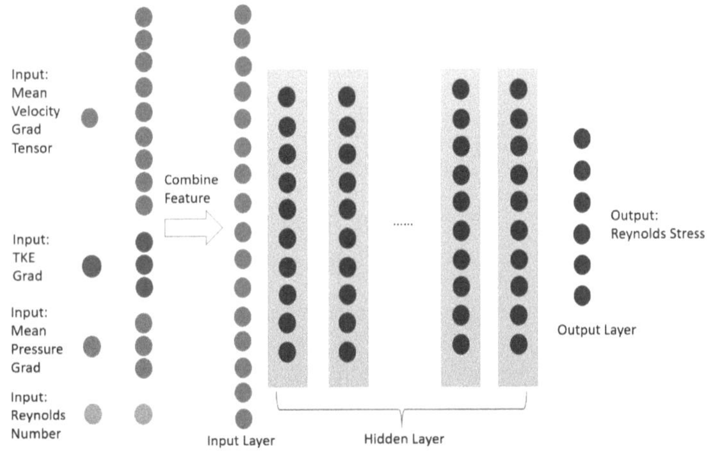

Fig. 3. FFNN-based turbulence modeling architecture designed from [8]

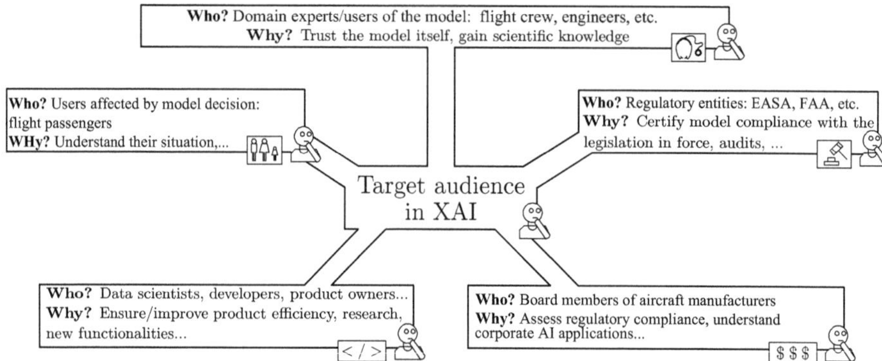

Fig. 4. Diagram showing the different purposes of explainability in the case of turbulence prediction following the audience profiles, adapted from [11], with permission from Elsevier.

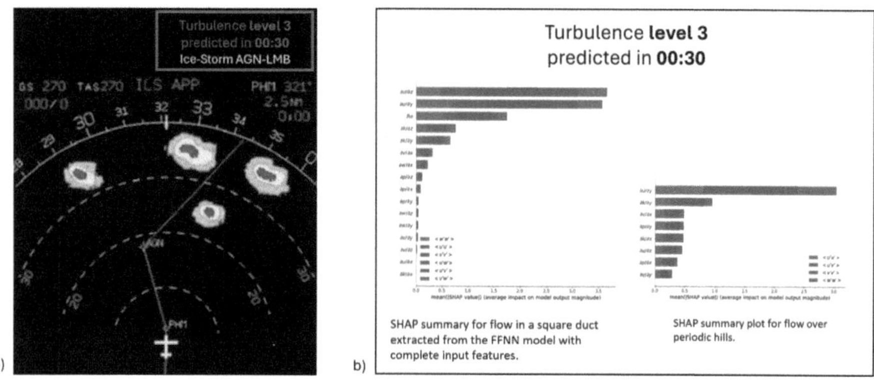

Fig. 5. Illustrative prototypes a) for the flight crew, and b) for the data scientists (adapted from [8]).

4.2   Opportunities

This example highlights opportunities aligned with those discussed in Sect. 3.

On the Engineering Side. The combination of a neural network with SHAP enabled the identification of influential input variables (e.g., components of the mean velocity gradient, Reynolds number, or turbulent kinetic energy gradient) that significantly affect Reynolds stress prediction. This contributes to feature discovery and highlights missing variables in classical turbulence models, offering insights for future model development.

On the End-User Side. Explanations derived from SHAP can increase trust by providing interpretable evidence of why the model predicts specific turbulence behaviours, and help understand the predictions better.

On the Certification Side. Explainability provides a basis for demonstrating compliance. For instance, SHAP values offer quantifiable evidence of how inputs contribute to outputs, which may support certification authorities in evaluating the model's robustness and fairness.

4.3   Challenges

Despite the opportunities presented by XAI, operationalizing it in critical systems such as turbulence prediction highlights several key challenges. These challenges can be grouped into three categories: usability, dependability, and confidentiality/privacy.

4.3.1   Usability Challenges

Task- and Role-Relevant Explanations. XAI explanations need to be tailored to the specific user role. Pilots require concise, operational explanations to support decision-making under time constraints, while data scientists and engineers require detailed, feature-level explanations to analyze and improve the model. Other stakeholders, such as regulators and manufacturers, have different informational needs that must also be considered (see Fig. 4).

As presented in Fig. 5, the explanations may be presented in different ways following the type of user. For instance, on the left of the Figure, the explanation presented to the flight crew gives the time to the turbulence, and the likely reason (an ice storm) supporting the tasks of operational decision-making (e.g., to decide to ask the air traffic controller to change the flight path or to decide to alert the crew and passengers). On the right of the Figure, the explanations provide much details to support the analysis tasks of the flight engineers and data scientists (e.g., to understand the features most involved in the prediction and to correct the model if necessary).

Temporal Constraints. Different users face different temporal constraints. Flight crews must interpret explanations in seconds, whereas engineers and data scientists can analyze explanations offline with no immediate time pressure. This requires adapting both the content and format of explanations to the decision-making timeframe of each user. The different types of users have different temporal constraints. This is illustrated by the example prototypes in Fig. 5: (i) The prototype on the left is designed for the flight crew, which has little time to analyze and interpret the explanations. Therefore, the explanation given to them is really succinct and oriented toward their needs: to know when the turbulence may occur (the predicted turbulence should happen in 30 s), to know where it is located (between waypoints AGN and LMB), and to know its nature (due to an ice-storm) to support the decision of actions to be taken. (ii) The prototype on the right is designed for the data scientists who are designing the model. These users do not have the same temporal constraints because they work on the model before it is deployed and have time to deeply analyze the explanations. Here, the prototype presents the detailed data of the involvement of all features in the final prediction.

Interface and Interaction Design. Interfaces must support navigation through explanations appropriate for each user role. Flight crews may receive static, succinct visualizations, while engineers require interactive tools to explore multiple explanation modalities (e.g., SHAP summary plots, feature importance charts). Progressive disclosure techniques, such as "Why?" buttons or layered explanations, can help manage information overload and guide users to relevant insights efficiently [22].

User Training. Embedding XAI requires adapting training programs. Pilots must learn to interpret operational explanations quickly and act accordingly. Engineers and data scientists must be trained to understand detailed, interactive explanations to refine models and detect errors. Training should account for each role's temporal and cognitive constraints. Both prototypes presented in Fig. 5 illustrate well the need for modifying users' training. Concerning the flight crew (prototype on the left), their training needs to be adapted to take into account this new data (predicted turbulence and corresponding explanation). Concerning the data scientists (prototype on the right), their training needs to be adapted to the potential interactions with these graphs.

4.3.2   Dependability Challenges

Reliability of Predictions. The reliability of AI explanations is fundamentally constrained by the data and scenarios on which the underlying model was trained. For example, as shown in [8], turbulence prediction models achieve high accuracy for specific flow configurations, such as square duct or periodic hill flows, but their generalization to other aerodynamic scenarios remains uncertain. This limitation introduces challenges for robustness and uncertainty management: explanations may be misleading when the model encounters out-of-distribution conditions, and users may over-rely on outputs without awareness of potential failure modes. Consequently, evaluating and communicating the confidence and limits of AI predictions is critical for decision-making in safety-critical contexts.

Evidence for Certification. In regulated domains, such as aviation, regulatory authorities require demonstrable evidence that both AI models and their explanations are correct, reproducible, and aligned with safety standards. Compliance with standards such as DO-178C [5] entails documenting the model's reasoning process and ensuring traceability of predictions. Operational XAI must therefore provide auditable reasoning paths that can be independently verified without compromising proprietary algorithms or sensitive data. This includes supporting systematic testing, uncertainty quantification, and scenario-based validation to facilitate certification and increase trust among stakeholders.

4.3.3   Confidentiality and Privacy Challenges

XAI solutions must generate explanations tailored to the needs and constraints of different user roles while respecting privacy and confidentiality. Incorporating role-based filtering, adhering to data minimization principles, and ensuring compliance with legal frameworks such as the GDPR [2] allows explanations to be both actionable and legally compliant without unnecessary disclosure of sensitive information.

In this illustrative example, explanations are generated at the feature level using SHAP across the full training dataset. While this provides useful insights for engineering analysis, it raises issues from a confidentiality and privacy perspective. Different levels of confidentiality and privacy need to be respected depending on the end user. Explanations expose patterns that are derived from all available training data, without any filtering or tailoring to specific end-user roles. In a real operational setting, such uniform disclosure could inadvertently reveal sensitive or proprietary information, and it does not support differentiated access control (e.g., pilots vs. engineers vs. regulators). Moreover, the approach does not incorporate mechanisms to ensure compliance with data protection frameworks such as the GDPR, where explanations should also respect the principles of data minimization and purpose limitation.

In concrete terms, pilots require explanations focused on operational impact (e.g., turbulence expected at flight level 340 within 10 min—consider descent to 320) without disclosure of raw meteorological datasets or proprietary prediction models. Maintenance engineers may need access to explanations at a deeper technical level, for instance, identifying which onboard sensors contributed most to the prediction, but still without exposure to sensitive passenger data or confidential third-party weather feeds. Regulators, in turn, may be provided with aggregated or abstracted explanations demonstrating compliance with safety requirements, while protecting both commercial confidentiality and personal data. The challenge lies in tailoring explanations to these roles without leaking sensitive technical details or confidential operational data.

5   Research Directions and Conclusion

This paper highlights several key challenges that must be addressed to effectively embed XAI within critical interactive systems. In light of these challenges, we identify several promising research directions.

First, there is a need for methods to identify and analyze how explanations can adapt to users' tasks, roles, and levels of expertise. A systematic analysis of users' goals and activities can facilitate this adaptation, for instance, by supporting the assessment of conformance and consistency between XAI outputs and user-centered tasks. Second, new evaluation methodologies are required to assess not only the correctness of explanations, but also their operational utility, usability, and impact on human–AI coordination and decision-making. Such evaluations should consider contextual performance metrics, including cognitive load, trust calibration, and resilience to system failures. Third, the integration of explainability into system certification and assurance processes remains largely unexplored. This raises important questions about the formal properties, verifiability, and traceability of explanation mechanisms, especially in regulated and safety-critical domains. Fourth, confidentiality and privacy aspects also have to be taken into account at design time.

Finally, another possible research direction is to explore adaptive and proactive explainability—mechanisms capable of anticipating information needs, detecting misunderstanding, and adjusting the level or form of explanation in real time. Such adaptive explainers could enhance both transparency and user engagement in dynamic, high-stakes environments.

These research directions call for a convergence of insights from XAI, human–computer interaction, and systems engineering. By operationalizing explainability as a context-sensitive, uncertainty-aware, and interactionally embedded property, we move toward AI systems that are not only intelligible but also meaningfully aligned with the usability, safety, and dependability requirements of critical interactive systems.

References

1. Easy Access Rules for Large Aeroplanes (CS-25) - Revision from June 2022. https://www.easa.europa.eu/en/document-library/easy-access-rules/easy-access-rules-large-aeroplanes-cs-25. Accessed 21 Sept 2025
2. General Data Protection Regulation (GDPR) – Legal Text. https://gdpr-info.eu/. Accessed 21 Sept 2025
3. ISO/DIS 9241-11:2018(en) ergonomics of human-system interaction — part 11: Usability: Definitions and concepts. https://www.iso.org/obp/ui/#iso:std:iso:9241:-11:ed-2:v1:en. Accessed 21 Sept 2025
4. ISO/DSI 9241-210:2019(en) ergonomics of human-system interaction - part 210: Human-centred design for interactive systems. https://www.iso.org/standard/77520.html. Accessed 21 Sept 2025
5. Software Considerations in Airborne Systems and Equipment Certification. https://standards.nasa.gov/standard/NASA/RTCA-DO-178. Accessed 21 Sept 2025
6. Aguinis, H., Kraiger, K.: Benefits of training and development for individuals and teams, organizations, and society. Annu. Rev. Psychol. **60**, 451–74 (2009). https://doi.org/10.1146/annurev.psych.60.110707.163505
7. Alecu, L., et al.: Can we reconcile safety objectives with machine learning performances? In: ERTS 2022 (2022)
8. Alhafiz, M.R., Zuhal, L.R., Dung, D.V., Palar, P.S.: An explainable deep learning for data-driven turbulence model feature discovery. IEEE Access **13**, 149801–149816 (2025). https://doi.org/10.1109/ACCESS.2025.3602021
9. Alzetta, F., Giorgini, P., Najjar, A., Schumacher, M.I., Calvaresi, D.: In-time explainability in multi-agent systems: challenges, opportunities, and roadmap. In: Calvaresi, D., Najjar, A., Winikoff, M., Främling, K. (eds.) EXTRAAMAS 2020. LNCS (LNAI), vol. 12175, pp. 39–53. Springer, Cham (2020). https://doi.org/10.1007/978-3-030-51924-7_3
10. Avizienis, A., Laprie, J.C., Randell, B., Landwehr, C.: Basic concepts and taxonomy of dependable and secure computing. IEEE Trans. Dependable Secure Comput. **1**(1), 11–33 (2004). https://doi.org/10.1109/TDSC.2004.2

11. Díaz-Rodríguez, N., et al.: Explainable artificial intelligence (XAI): concepts, taxonomies, opportunities and challenges toward responsible AI. Inf. Fusi. **58**, 82–115 (2020). https://doi.org/10.1016/j.inffus.2019.12.012
12. Bilgic, M., Mooney, R.J.: Explaining recommendations: satisfaction vs. promotion. In: Beyond Personalization Workshop, IUI, vol. 5, p. 153 (2005)
13. Bouzekri, E., et al.: Engineering issues related to the development of a recommender system in a critical context: application to interactive cockpits. Int. J. Hum. Comput. Stud. **121**, 122–141 (2019). Advances in Computer-Human Interaction for Recommender Systems. https://doi.org/10.1016/j.ijhcs.2018.05.001
14. Bouzekri, E., Martinie, C., Palanque, P., Atwood, K., Gris, C.: Should i add recommendations to my warning system? The RCRAFT framework can answer this and other questions about supporting the assessment of automation designs. In: Ardito, C., et al. (eds.) INTERACT 2021. LNCS, vol. 12935, pp. 405–429. Springer, Cham (2021). https://doi.org/10.1007/978-3-030-85610-6_24
15. Chromik, M., Butz, A.: Human-XAI interaction: a review and design principles for explanation user interfaces. In: Ardito, C., et al. (eds.) INTERACT 2021. LNCS, vol. 12933, pp. 619–640. Springer, Cham (2021). https://doi.org/10.1007/978-3-030-85616-8_36
16. Delaunay, J., Largouët, C., Galárraga, L., Van Berkel, N.: Adaptation of AI explanations to users' roles. In: Workshop on Human-Centered Explainable AI, HCXAI (2023). https://inria.hal.science/hal-04388942/
17. Delseny, H., Gabreau, C., Gauffriau, A., Beaudouin, B., Ponsolle, L., et al.: White paper machine learning in certified systems (2021). https://doi.org/10.48550/ARXIV.2103.10529
18. Došilović, F.K., Brčić, M., Hlupić, N.: Explainable artificial intelligence: a survey. In: 2018 41st International Convention on Information and Communication Technology, Electronics and Microelectronics (MIPRO), pp. 0210–0215 (2018). https://doi.org/10.23919/MIPRO.2018.8400040
19. Felzmann, H., Villaronga, E.F., Lutz, C., Tamò-Larrieux, A.: Transparency you can trust: Transparency requirements for artificial intelligence between legal norms and contextual concerns. Big Data Soc. **6**(1), 2053951719860542 (2019). https://doi.org/10.1177/2053951719860542
20. Garouani, M., Ahmad, A., Bouneffa, M., Hamlich, M., Bourguin, G., Lewandowski, A.: Towards big industrial data mining through explainable automated machine learning. Int. J. Adv. Manuf. Technol. **120**(1–2), 1169–1188 (2022). https://doi.org/10.1007/s00170-022-08761-9
21. Garouani, M., Barhrhouj, A., Teste, O.: XStacking: an effective and inherently explainable framework for stacked ensemble learning. Inf. Fus. **124**, 103358 (2025). https://doi.org/10.1016/j.inffus.2025.103358
22. Garouani, M., Bouneffa, M.: Unlocking the black box: towards interactive explainable automated machine learning. In: Quaresma, P., Camacho, D., Yin, H., Gonçalves, T., Julian, V., Tallón-Ballesteros, A.J. (eds.) Intelligent Data Engineering and Automated Learning, IDEAL 2023. LNCS, vol. 14404, pp. 458–469. Springer, Cham (2023). https://doi.org/10.1007/978-3-031-48232-8_42
23. Garouani, M., Mothe, J., Barhrhouj, A., Aligon, J.: Investigating the duality of interpretability and explainability in machine learning. In: 2024 IEEE 36th International Conference on Tools with Artificial Intelligence (ICTAI), pp. 861–867 (2024). https://doi.org/10.1109/ICTAI62512.2024.00125
24. Gedikli, F., Jannach, D., Ge, M.: How should i explain? A comparison of different explanation types for recommender systems. Int. J. Hum Comput Stud. **72**(4), 367–382 (2014). https://doi.org/10.1016/j.ijhcs.2013.12.007

25. Haid, C., Lang, A., Fottner, J.: Explaining algorithmic decisions: design guidelines for explanations in user interfaces. Hum. Fact. Softw. Syst. Eng. **94**(94) (2023). https://doi.org/10.54941/ahfe1003764
26. Herlocker, J.L., Konstan, J.A., Riedl, J.: Explaining collaborative filtering recommendations. In: Proceedings of the 2000 ACM Conference on Computer Supported Cooperative Work, pp. 241–250 (2000). https://doi.org/10.1145/358916.358995
27. Li, T., Goupil, P., Mothe, J., Teste, O.: Real-time early identification of atmospheric turbulence using flight sensor data. J. Air Transp., 1–12 (2025). https://doi.org/10.2514/1.D0391
28. Martinie, C.: Challenges for operationalizing XAI in critical interactive systems. In: ACM CHI Workshop on Operationalizing Human-Centered Perspectives in Explainable AI (HCXAI@ CHI 2021) (2021)
29. Millecamp, M., Htun, N.N., Conati, C., Verbert, K.: To explain or not to explain: the effects of personal characteristics when explaining music recommendations. In: Proceedings of the 24th International Conference on Intelligent User Interfaces, pp. 397–407 (2019). https://doi.org/10.1145/3301275.3302313
30. Nimmo, R., Constantinides, M., Zhou, K., Quercia, D., Stumpf, S.: User characteristics in explainable AI: the rabbit hole of personalization? In: Proceedings of the 2024 CHI Conference on Human Factors in Computing Systems, CHI '24. Association for Computing Machinery, New York, NY, USA (2024). https://doi.org/10.1145/3613904.3642352
31. Raji, I.D., et al.: Closing the ai accountability gap: defining an end-to-end framework for internal algorithmic auditing. In: Proceedings of the 2020 Conference on Fairness, Accountability, and Transparency, FAT* '20, pp. 33–44. Association for Computing Machinery (2020). https://doi.org/10.1145/3351095.3372873
32. Ribera, M., Lapedriza, A.: Can we do better explanations? A proposal of user-centered explainable ai. In: Proceedings of the ACM IUI 2019 Workshops (2019)
33. Scholes, M.S.: Artificial intelligence and uncertainty. Risk Sci. **1**, 100004 (2025). https://doi.org/10.1016/j.risk.2024.100004
34. Springer, A., Whittaker, S.: Progressive disclosure: When, why, and how do users want algorithmic transparency information? ACM Trans. Interact. Intell. Syst. (TiiS) **10**(4), 1–32 (2020). https://doi.org/10.1145/3374218
35. Vössing, M., Kühl, N., Lind, M., Satzger, G.: Designing transparency for effective human-AI collaboration. Inf. Syst. Front. **24**(3), 877–895 (2022). https://doi.org/10.1007/s10796-022-10284-3
36. Wachter, S., Mittelstadt, B., Russell, C.: Counterfactual explanations without opening the black box: automated decisions and the GDPR (2018). https://arxiv.org/abs/1711.00399
37. Weber, L., Lapuschkin, S., Binder, A., Samek, W.: Beyond explaining: opportunities and challenges of XAI-based model improvement. Inf. Fus. **92**, 154–176 (2023). https://doi.org/10.1016/j.inffus.2022.11.013
38. Yu, B., Yuan, Y., Terveen, L., Wu, Z.S., Forlizzi, J., Zhu, H.: Keeping designers in the loop: communicating inherent algorithmic trade-offs across multiple objectives. In: Proceedings of the 2020 ACM Designing Interactive Systems Conference, pp. 1245–1257 (2020). https://doi.org/10.1145/3357236.3395528

Towards a Unified User Modeling Language for Engineering Human Centered AI Systems

Aaron Conrardy[1,2]([envelope]) [ORCID], Alfredo Capozucca[2] [ORCID], and Jordi Cabot[1,2] [ORCID]

[1] Luxembourg Institute of Science and Technology, Esch-sur-Alzette, Luxembourg
{aaron.conrardy,jordi.cabot}@list.lu
[2] University of Luxembourg, Esch-sur-Alzette, Luxembourg
alfredo.capozucca@uni.lu

Abstract. In today's digital society, personalization has become a crucial aspect of software applications, significantly impacting user experience and engagement. A new wave of intelligent user interfaces, such as AI-based conversational agents, has the potential to enable such personalization beyond what other types of interfaces could offer in the past. Personalization requires the ability to specify a complete user profile, covering as many dimensions as possible, such as potential accessibility constraints, interaction preferences, and even hobbies. Yet, existing solutions for user modeling mostly focus on individual aspects at a very coarse level, severely limiting the potential adaptations for personalization. In this sense, this paper presents a unified user modeling language, aimed to combine previous approaches, both from the modeling community and other user-centric fields, in a single proposal. This language has been implemented on top of the open source BESSER low-code platform. Additionally, a proof of concept leveraging user profiles modeled with our language to automatically adapt a conversational agent has also been developed.

Keywords: User Modeling Language · User Model · Model-Driven Engineering · Personalization · User Profile

1 Introduction

Users are at the center of interactive applications [7]. Yet, each user is different and has different needs and expectations based on their specific profile (age, mood, personality, etc.). Therefore, to provide an optimal user experience, applications need to adapt to each specific profile, providing a personalized interaction. Failing to do so could even lead to digital inequalities, as certain groups of users may not able to fully take advantage of an application due to their specific limitations (e.g. accessibility issues or language skills) [26].

Moreover, the growth of AI has increased the possibilities of user profiling and personalizing user interactions. Examples such as inferring the user's fatigue level based on smartwatch data using machine learning (ML) models [16] or predicting a user's behavior and attitudes based on an interview using Large Language Models (LLMs) [21] have now become possible. Furthermore, based on this profile data, AI also increases the degree of personalization options [23], such as automatically tuning text based on the language skills of the user [20] or performing user interface layout optimization based on user data [11].

To enable all these new opportunities, we need to be able to model such user profiles in a structured and machine-readable manner, which is known as user modeling [25]. Nevertheless, there is not yet a general user modeling language [2,15] that could be used as the *de facto* standard for the model-driven engineering (MDE) community and beyond. Conversely, there are myriad of partial and overlapping solutions that result in incomplete profiles and interoperability issues.

This paper aims to provide such user modeling language by collecting and unifying previous proposals from the modeling community, complemented with user profiling perspectives from other fields (sociology, psychology, etc.). Additionally, to illustrate the benefits of our language for personalization purposes, we have implemented a proof of concept that automatically adapts the behavior of an LLM agent based on a given user profile.

The rest of the paper is structured as follows: Sect. 2 describes the state of the art on user modeling languages and current drawbacks. Section 3 presents the metamodel of the proposed unified user modeling language and discusses a possible concrete syntax. The implemented proof of concept is described in Sect. 4 and the corresponding tool support in Sect. 5. Finally, we present future work in Sect. 6 and conclude the paper in Sect. 7.

2   State of the Art

We have conducted a systematic literature review (SLR) on user modeling in MDE [2]. The SLR comprised 30 papers that proposed at least some kind of abstract formalization of a user modeling language (such as a metamodel or grammar) and investigated its integration into engineering pipelines. In what follows, we briefly discuss the most relevant results.

As stated before, there is no unified user modeling language. Instead, most works rather attempted to create a new proposal from scratch, ignoring existing results, and thus, leading to a fragmented community that proposes overlapping user dimensions while, at the same time, missing others already proposed in previous works. Additionally, there is a tendency to favor more static user aspects, while neglecting more complex dynamic ones (like mood or fatigue) and/or present the dimensions at a very high level without giving enough details on the possible values that could be set for each dimension. A possible explanation for these limitations is that, in most of such works, modeling the user is a secondary concern rather than a first-class citizen, missing out on potential personalization opportunities.

Out of the reviewed papers, there were two exceptions that had a goal similar to ours. Kaklanis et al. [12] aimed to provide a standard of user models for the purpose of user simulation in virtual environments. They explicitly mention that their focus was on an interoperable user model to describe able-bodied individuals and people with various kinds of disabilities. Thus, the scope is rather narrow compared with ours. Similarly, Gaspar et al. [6] created a unified user model, this time specifically for dimensions that directly affect the user interface to be presented to the user. Again, our target language aims to be more generic and enable all types of user adaptations. Even more importantly, the dimensions defined by Gaspar et al. are not detailed and stay at a very generic level which hinders the reusability of this approach.

3   A Unified User Modeling Language

To overcome the fragmented nature of user models and provide a more complete, while at the same time, fine-grained, representation of users, we developed a new language that integrates insights from the SLR and user dimensions from other fields. This section presents the abstract and concrete syntax of the unified user modeling language.

3.1   Abstract Syntax: The Unified User Metamodel

We begin by defining the abstract syntax of the user modeling language, formalized as a metamodel and represented, as usual, using the UML class diagram notation. As starting point for the metamodel, we took the results from the conducted SLR [2] as it provided a list of user dimensions and a classification for these. We have then extended such list of dimensions, e.g. by proposing the additional category "Culture", as it is regarded as another important source of influence of the user [7,24], but was missing from the reviewed literature. In some cases, dimensions are shared by multiple categories (e.g. age fits in both "Personal Information" and "Accessibility"), yet we present them as part of the most fitting one for convenience.

Figure 1 depicts a high-level view of the metamodel, showcasing the main categories that characterize a user. Note that all categories (and the dimensions within) are optional as users may decide not to disclose that information or, in automatic profiling methods, we may not know them from the beginning.

Furthermore, as many dimensions include a range value, we introduce a new datatype with the same name to represent integer numbers in the range of 0 to 100. In most cases, this datatype is used when a dimension represents a score, such as the level of happiness a user is currently experiencing. Some dimensions instead compare two opposing concepts. In those cases, the Range is used as a scale following the naming scheme "X_To_Y" to compare X to Y, such as "bad_to_good" (Fig. 7) that represents whether a user is leaning towards a bad (value leaning towards 0) or good (value leaning towards 100) mood. Range generally offers flexibility, as depending on the requirements, one could map a predefined selection of choices to the values of Range (e.g. a binary selection of

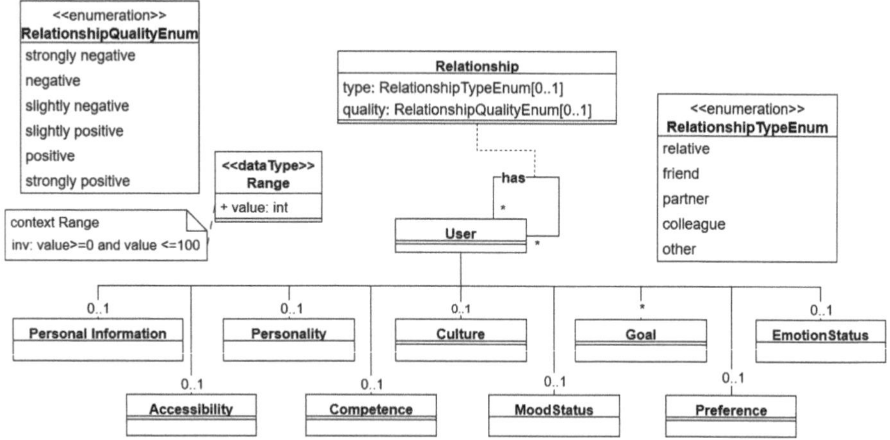

Fig. 1. High-level view of unified user metamodel

"No/Yes" could be mapped to the values 0 and 100 of Range), simulating an enumeration without altering the metamodel.

In the following, we will go over the main concepts of the metamodel, including all categories and briefly discuss their content. Due to size limitations, parts of the different metamodel are presented condensed. A complete version is available online.[1] The online version contains the complete metamodel, detailed information on the user dimensions and a mapping of reference to user dimension.

Relationship. Information on user relationships with other users is popular in research works that deal with user models for social network platforms (e.g. [13]). However, most models only provide information on the existence of a relationship but not the type of relationship or any other relevant data. We propose the possibility to also capture the type of relationships (relative, partner, colleague, etc.) and the quality of the relationship (Relationship class in Fig. 1). This information is valuable in understanding the nuance of a relationship, as not every relationship is of positive nature and it may affect processes such as content recommendation.

Personal Information. Personal information (Fig. 2) describes typical information a user would enter on a profile page when registering for a new social application (e.g. a social network account). That includes information such as the full name, age, address or gender of the user.

Regarding "Interest" and "Hobby", while both deal with engaging in topics via an activity, "Hobby" tackles a more recurrent and dedicated engagement. Therefore, we model "Hobby" as a stronger version of "Interest" where we indicate the number of weekly hours the user devotes to the hobby. The "Topic" class

---

[1] https://github.com/BESSER-PEARL/User-Modeling-Language.

represents a non-exhaustive list of topics, derived from an ontology provided in
[8].

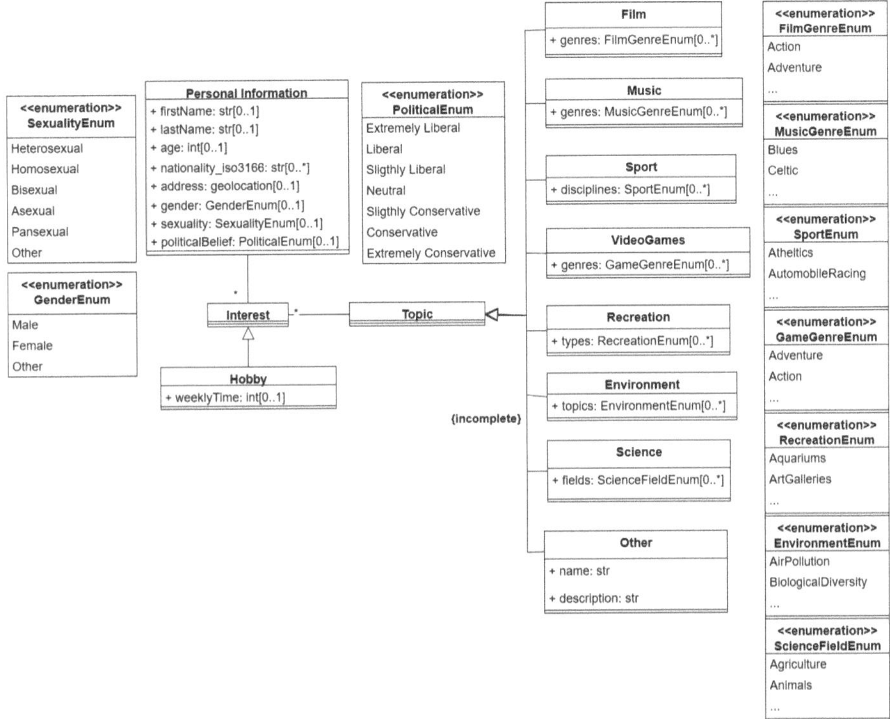

Fig. 2. Personal information

Competence. Competence (Fig. 3) describes a user's proficiency on a number of aspects such as education, language or a various number of skills and expertises. Regarding languages, we propose the usage of the Common European Framework of Reference for Languages[2] (CEFR) as a standard description of a user's language skills. Regarding "Skill" and "Knowledge", the attribute "score" is used to describe the user's ability to perform a skill or level of knowledge respectively. For the latter, "Knowledge" relates to the topics contained in the previously presented "Topic" class.

Accessibility. Accessibility (Fig. 4) refers to the accessibility needs of the user. The "Disability" class describes a concrete disability a user might have, whereas

---

[2] https://www.coe.int/en/web/common-european-framework-reference-languages/
level-descriptions.

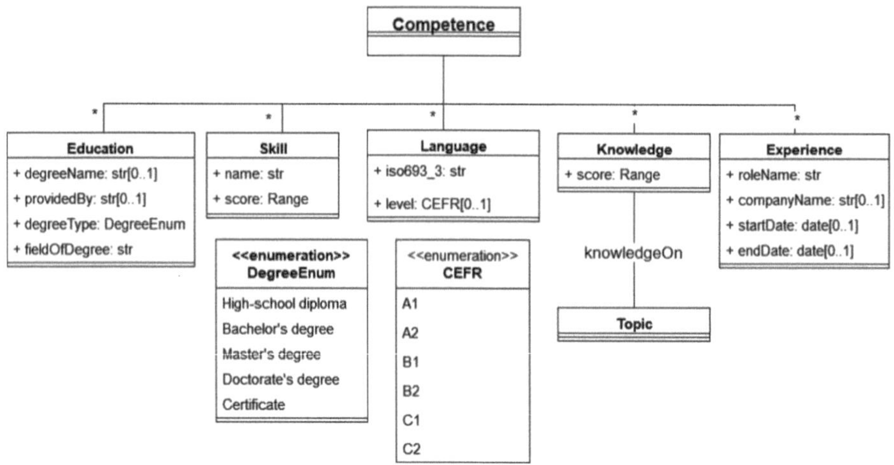

Fig. 3. Competence

the other classes ("Mobility", "Speech", "Hearing", etc.) describe the state of the different senses, parts of the body and body functions. Of course, Disability and the other classes affect each other, as a completely deaf person will have the lowest hearing values. While most of the classes are inspired by Kaklanis et al. [12], their metamodel did not contain any attribute types. Therefore, we completed the metamodel by providing attribute types based on the used values from the results of common measurement methods for a bodily function. An example would be the usage of decibels to measure a person's hearing during an audiogram [22], making the choice of integers fitting.

Personality. Personality (Fig. 5) tackles the user's personality traits, that is, the stable internal characteristics of the users. Numerous personality models were developed by psychologists to explain the nature of humans and their differences. For our purposes, we opted to cover personality traits from the most popular models (as proposed in [8]), such as the Big Five personality traits [4]. These personality traits, represented by the "Trait" class, describe people's pattern of thoughts, feelings and behaviors [3], while the "Characteristic" class describes observable and concrete feature of a person implied by a trait. Both classes have a score to denote the extent to which a user fulfills a trait/characteristic.

Additional dimensions affected by, or linked to personality were included, such as Motivation and Attitude. The former is based on a unified motivational model proposed by Forbes [5] that explains the different motivational constructs a person might lean towards to. The latter tackles a person's opinion given a topic, be it of a negative or positive nature. Bias in that sense is a stronger and more negative version of Attitude, often leading to an unfair treatment of a target.

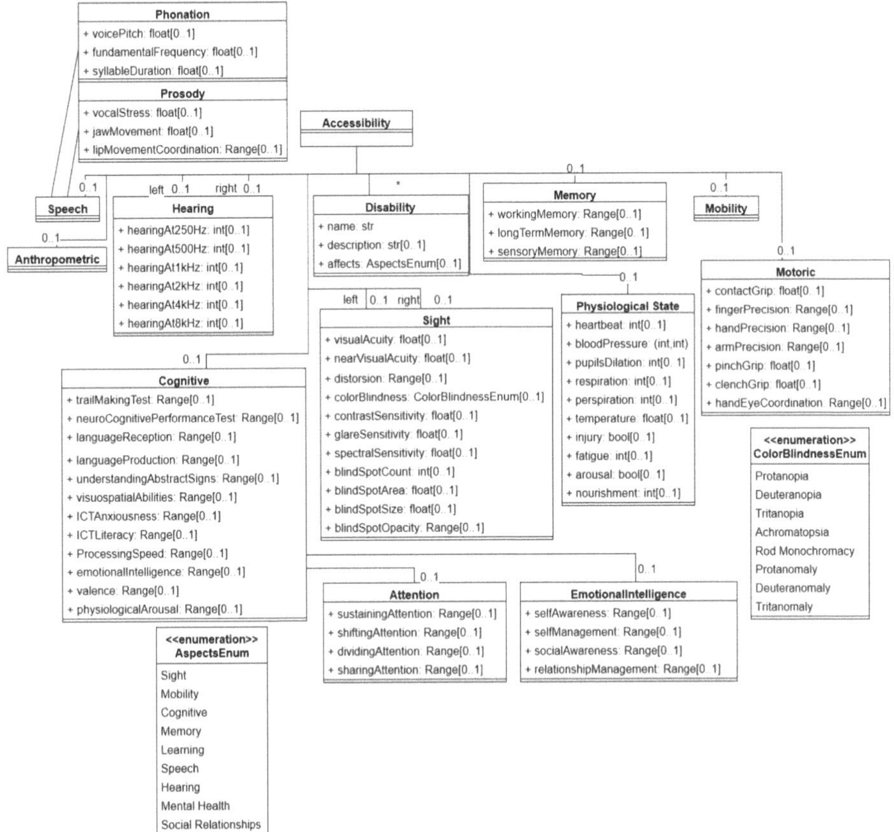

Fig. 4. Accessibility

Preference. Preferences (Fig. 6) are used to represent a user's personal preferences, that is, a preferred option in regards to a set of choices. In the context of a software application, the choices are narrowed to preferences of the content itself (such as preferring one topic over another or the preferred language), design preferences (such as a preferred color scheme) and the preferred interaction modality (such as audio or textual interaction).

Culture. In general, culture (Fig. 7) is linked to the social behavior, institutions and norms found in human societies and is often attributed to a specific region/location. To model the culture of a user, we followed Hofstede's cultural dimensions [10]. These consist of dimension pairs that are opposed to each other (e.g. Collectivism and Individualism), which is why we used our Range datatype. Additionally, we added Religion as a dimension, as it is often linked to the culture of regions/locations and influences expectations during user interaction with an application [18].

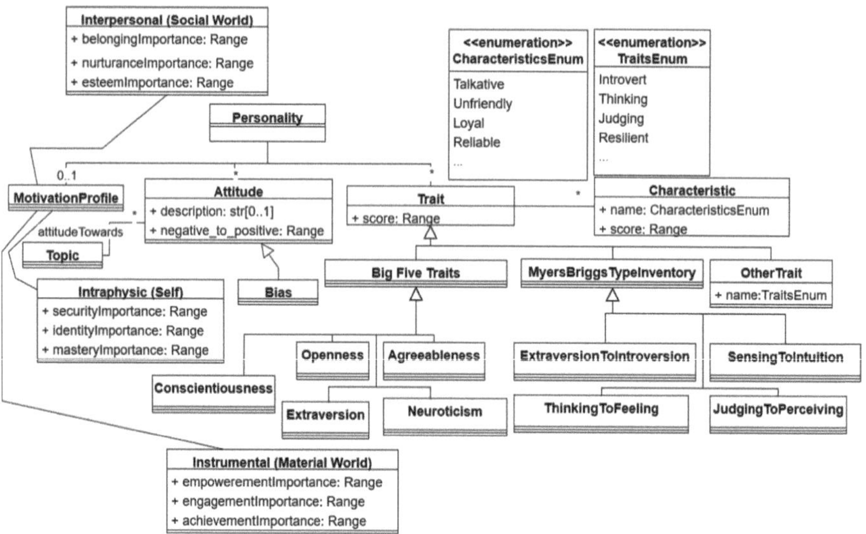

Fig. 5. Personality

Goals. Goals (Fig. 7) focus on the objectives of a user, both while interacting with an application and outside. Optionally, a goal can have a deadline. Essentially, goals provide a purpose behind the user's actions and can be both static or dynamic depending on the circumstances (a goal could appear based on an interaction that just took place).

Emotions and Moods. Similarly to personality, there are various models in psychology that attempt to quantify emotions and moods (Fig. 7). Regarding the concrete difference between emotions and moods, emotions tend to be associated with a specific object or event and do not last long, while moods tend to last longer and have a less direct cause [14]. For emotions, we decided to showcase the list of emotions as presented in GUMO [9]. For moods, we opted for the three pairs of opposing moods as presented in [17].

3.2   Concrete Syntax

Based on the presented abstract syntax, we derived a textual concrete syntax to facilitate the creation of user models. In particular, we have chosen a JSON-based notation due to its popularity and its human- and machine-readable nature, which facilitates the use of user models in AI-driven personalization scenarios. Indeed, LLMs were already shown to be able to understand [27] as well as produce JSON [19]. Beyond LLMs, many tools work with JSON documents which opens the door to numerous applications exploiting user models.

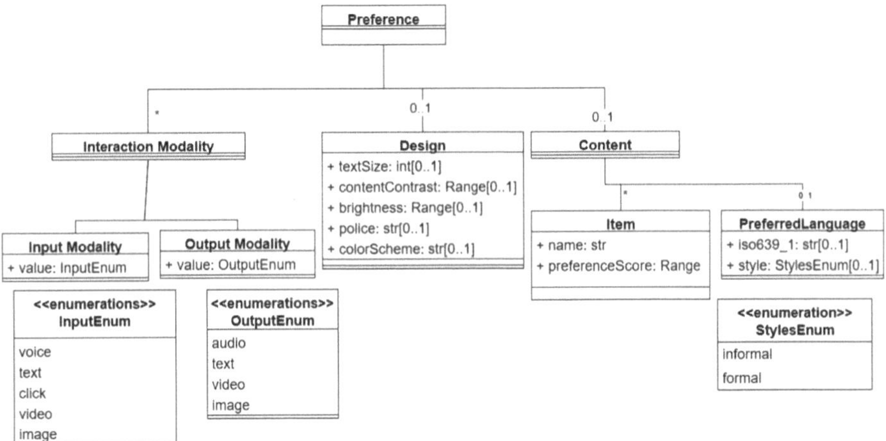

Fig. 6. Preference

JSON additionally allows for the definition of a JSON schema[3] that enables data consistency and validity checks of given JSON data based on predefined rules. In our case, the JSON Schema defines the concepts in the metamodel, and user models are represented as JSON objects that conform to this schema. Figure 8a shows an excerpt of the metamodel in JSON schema format whereas Fig. 8b showcases the use of the concrete syntax to write user profile data that can be validated using the defined schema.

In Sect. 5, we demonstrate how JSON objects representing the user model can be automatically generated from a form-based web application, further simplifying their creation.

4 Proof of Concept: Automatic Conversational Agent Personalization

To showcase the usefulness of the user modeling language, we have implemented a personalization pipeline that, given an input user model, automatically personalizes the responses of a conversational agent to increase the user experience and engagement.

Our pipeline supports two approaches to personalize conversational agents (depicted in Fig. 9), depending on the capabilities and access we have to the target LLM powering the conversational agent:

– Direct personalization of LLM-agent: The information of the user model is directly embedded into the LLM via fitting context prompts. As a result, the conversation responses generated by the LLM are inherently personalized.

---

[3] https://json-schema.org/.

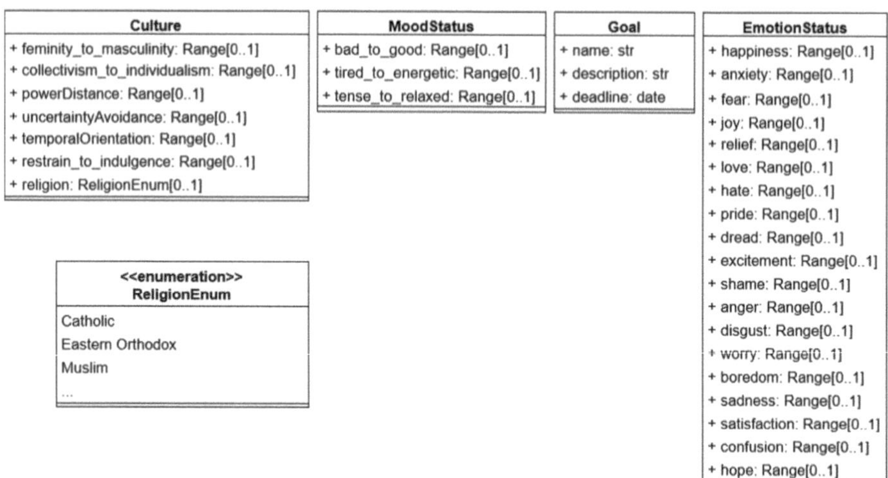

Fig. 7. Culture, Mood, Goal and Emotion

– Indirect personalization of the conversational agent: When we have an non-personalized message, whether it comes from the LLM without user context or is pre-defined by a developer, we post-process it. We do this by sending the original message and the user model to a secondary LLM, which produces a personalized version.

In the following, we describe the first approach. The process of the second approach is similar, except that it involves a separate request to an LLM with a slightly different prompt. We propose this second approach for two reasons: (1) it allows using one LLM optimized for text generation and another optimized for personalization, and (2) it supports a closed conversational agent with pre-defined conversation paths, preventing the user from engaging in open-ended dialogue.

Initially, a web application provides an interface for end-users to interact with a non-personalized LLM, with the option to upload a user model for personalization. Once a user model that follows the JSON schema as described in Sect. 3.2 is uploaded, the following prompt context is automatically added to the conversation request "You adapt your responses based on a given user profile, such as their native language, interests, and other provided attributes. Your goal is to enhance the user's experience by tailoring your responses to their profile. This is the user's profile {JSON file containing User Model}". As mentioned before, we take advantage of LLM's capabilities to understand JSON and therefore omit the transformation of the user model to a natural language form.

Once this sub-prompt is ready, it is automatically embedded into an LLM's context information. In our case, we used OpenAI's[4] GPT-4o-mini model due to its current reputation as one of the most performant LLM [28]. Using OpenAI's

---

[4] https://openai.com/.

```
{"$schema": "http://json-schema.org/draft-07/schema#",
 "title": "Generated JSON Schema",
 "type": "object",
 "properties": {
   "User": {
       "type": "object",
       "properties": {
         "PersonalInformation": {
           "type": "object",
           "properties": {
             "nationality": {
               "type": "array",
               "items": {
                 "$ref": "#/definitions/NationalityEnum"
               }
             },
             "age": {
               "type": "integer"
             }
           }
         },
         "Accessibility": {
           "type": "object",
           "properties": {
             "Disability": {
               "type": "array",
               "items": {
                 "properties": {
                   "name": {
                     "type": "string"
                   },
                   "description": {
                     "type": "string"
                   },
                   "affects": {
                     "$ref": "#/definitions/AspectsEnum"
                   }
                 }
               }
             }
           }
         }
       }
   }
 },
 "definitions": {
     "AspectsEnum": {
         "type": "string",
         "enum": [
             "Mobility","Sight"
         ]
     },
     "NationalityEnum": [...]
 }
}
```

(a) JSON schema

```
"User": {
    "PersonalInformation": {
      "age": 80,
      "nationality": ["Luxembourg"]
    },
    "Accessibility": {
      "Disability": [
        {
          "name": "Paraplegic",
          "description": "Unable to move legs",
          "affects": "Mobility"
        }
      ]
    }
}
```

(b) Instantiation following concrete syntax

Fig. 8. Example use of JSON syntax following the JSON schema

API[5], one can set context information at the system level, leading to an improved upholding of defined context rules.

Finally, the web application updates its layout to indicate that the user is now interacting with a personalized LLM.

As an example, Fig. 10 shows the initial web application and an example interaction with GPT-4o-mini, whereas Figs. 11a and 11b contain a comparison of the same question being asked with different user models stored in its context. Specifically, we prompted the LLM to generate exercise recommendations for muscle gain, constrained to three sentences, reflecting the use case of an AI coach providing beginner-friendly gym advice while considering the user's profile. We notice an adapted response based on the user's age or disability, showcasing the LLM's capabilities to adapt its responses based on the given combination of prompt and user model. We argue that in this simple example, the personalized messages improve the user experience for the given profiles. Naturally, a user

---

[5] https://platform.openai.com/docs/overview.

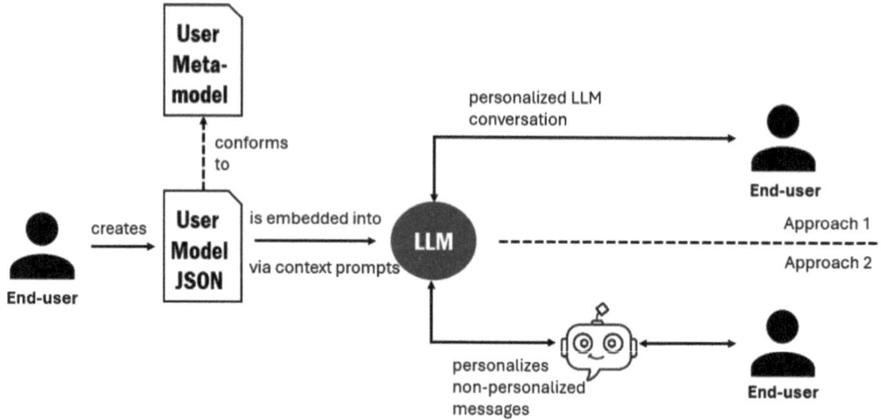

Fig. 9. Implemented approaches to personalize conversational agent

study would be necessary to confirm this. Such a study could consist of showing the default and personalized responses and letting participants evaluate their preferred answer.

5   Tool Support

The user modeling language and the proof of concept were implemented using existing and extended components from the open-source BESSER[6] low-code platform [1] and are both available online[7].

For the user modeling language, we first used BESSER's existing modeling language called B-UML to model the abstract syntax as a structural model following its textual syntax. We developed a new generator on top of the BESSER generators that performs a model to text transformation, transforming an input structural model into the equivalent JSON schema as defined in Sect. 3.2. The advantage of this generator is the possibility to easily update the abstract syntax and immediately updating the JSON schema. This property enables future contributors to easily extend the user modeling language.

To facilitate the usage of the concrete syntax, we developed a Streamlit[8] web-based application that lets end-users create a user model by simply filling a form, also available on the repository of the user modeling language. Following the defined cardinalities, a user has the option to only fill out a part of the form and can leave some information out. Figure 12 shows an excerpt of the personal information forms. Once filled out, the information is mapped to the concrete syntax and stored as a JSON file.

---

[6] https://github.com/BESSER-PEARL/BESSER.
[7] https://github.com/BESSER-PEARL/User-Modeling-Language.
[8] https://streamlit.io/.

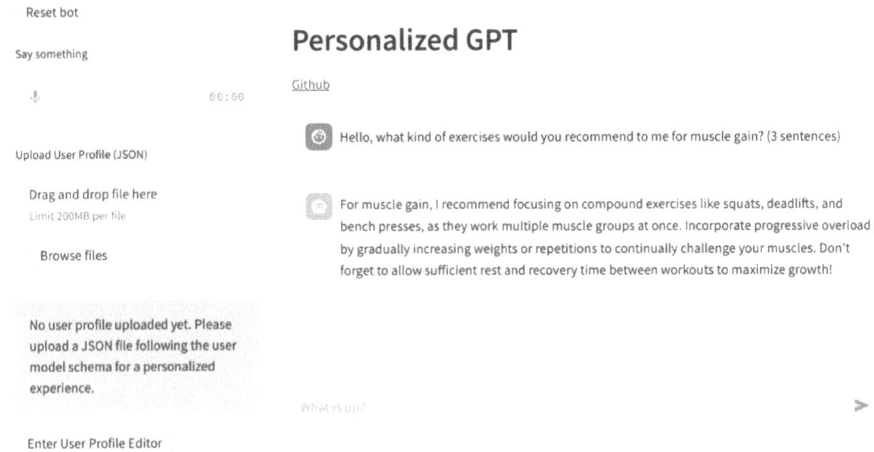

Fig. 10. Overview of chat interface and example conversation with LLM (GPT-4o-mini) without personalization

The personalizable LLM was developed using BESSER's agentic framework[9]. The framework contains wrappers for different LLMs and allows for the definition of context information that will be provided to the LLMs as high-priority context information. Additionally, BESSER proposes the previously shown frontend (Fig. 10) to interact with the created personalized LLM-agents. Both the forms and conversational agents frontend are combined, such that one can enter the forms page via the conversational agent frontend ("User Profile Editor" button in Fig. 10), enter the profile information and start the conversation with the personalized agent, removing the need for the end-user to manually import the user model.

6   Future Work

Interleaving with Existing Application Modeling Languages to Engineer AI Systems. The proposed proof of concept does not allow to define which kind of adaptation takes place, but rather, acts non-deterministically and performs automatic adaptation outside of our control. The defined user modeling language should be combined with application modeling languages (e.g. IFML[10]), specifying different kinds of behaviors based on specific values. The resulting combined language should allow for specifying both automatic and predefined adaptations, taking the best of both worlds. Existing attempts at combining user profile information and application behavior, such as [29], could act

---

[9] https://github.com/BESSER-PEARL/BESSER-Agentic-Framework.
[10] https://www.ifml.org/.

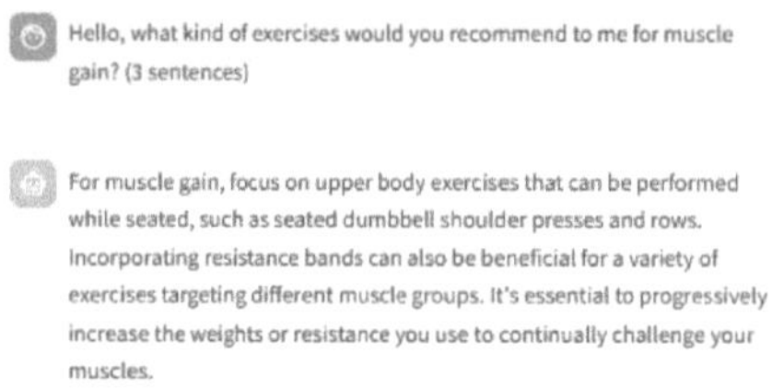

(a) Profile of paraplegic male aged 30: We notice the omission of exercises involving the lower body and an increased focus on upper body exercises

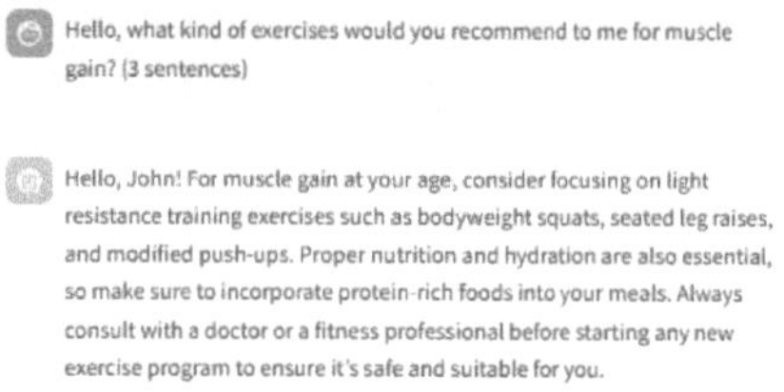

(b) Profile of male aged 80: We notice a focus on light exercises, avoiding weighted exercises and recommendations to visit a doctor to be safe due to the age

Fig. 11. Interaction with personalized LLM (GPT-4o-mini) using different user models

as an inspiration and be extended, avoiding re-inventing the wheel. Correspondingly, generators shall be implemented that ingest the models and produce the adaptive application, enabling a pipeline that allows developers to both design and automatically produce (parts of) a user centered AI system.

Additionally, each language should be equipped with a graphical concrete syntax, increasing the user experience for the developers, providing straightforward drag and drop components leading to an easy and fast development.

Quality Evaluation of User Models. Performing quality evaluations of created user models using model-based testing, verification and validation approaches is mostly unexplored [2]. Existing approaches used for application models could be adapted to be applied for user models. Techniques to check the user model's consistency (e.g. a deaf user should not declare audio as a preferred modality) or verifying that pre-defined rules are evaluated to true (e.g. the age of a user cannot be over 200) should be included in the user modeling language.

Automatic Profiling of User. Beyond static information to be provided by the end-user, we will explore how to support dynamic scenarios where applications need to adapt to users for which there is no profile information available. For such users the idea would be to discover a basic profile through the interactions with the systems to, slowly but steadily, adapt the interface to their profile. Existing methods (e.g. emotion recognition using [29] computer vision) and emerging ones (e.g. inferring the user's behaviors and attitudes based on a conversation with an LLM-agent [21]) need to be explored to allow the application to create an accurate model of the user via explicit and implicit methods.

Ethical and Privacy Considerations. The processing of personal data is governed by legal and ethical requirements, most notably the General Data Protection Regulation (GDPR). Consequently, any personalization framework must

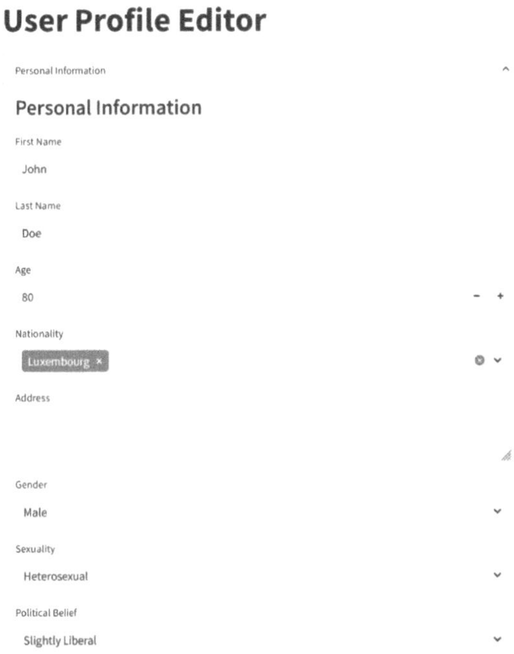

Fig. 12. Forms frontend

be designed to ensure compliance with these requirements. Key aspects include obtaining and managing user consent, minimizing the collection of unnecessary data, limiting data retention to what is strictly required for processing, and pseudonymizing data when delegating processing to third parties.

7   Conclusion

In this paper, we have proposed a unified user modeling language that is the sum and adaptation of fragmented solutions from the MDE community and beyond. First, the abstract syntax was formalized as a metamodel that covers a wide range of clearly defined user dimensions, allowing for a complete representation of any user and increasing the personalization potential. Secondly, the concrete syntax is based on a JSON schema that was derived from the metamodel and enables the creation of user models that capture the users' information. Compared with the reviewed works [2], our language is much more complete, as the maximum number of categories (respectively user dimensions) covered by a solution in previous works was 5 (respectively 14), while our metamodel covers 10 categories and provides more granular details on the proposed dimensions, clearly providing a more exhaustive palette to represent users.

Moreover, the proof of concept showcases the usefulness of the user modeling language by using the produced user models to automatically personalize a conversational agent. We argue that the provided adaptation improves the quality of the response for the given profile and, consequently, the overall user experience.

Implementing the points defined in future work will establish a pipeline to model, test and generate human centered AI systems, speeding up their development while ensuring compliance with existing regulations.

Acknowledgments. This project is supported by the Luxembourg National Research Fund (FNR) PEARL program, grant agreement 16544475.

References

1. Alfonso, I. et al.: Building BESSER: an open-source low-code platform. In: van der Aa, H., Bork, D., Schmidt, R., Sturm, A. (eds.) Enterprise, Business-Process and Information Systems Modeling, BPMDS EMMSAD 2024. LNBIP, vol. 511, pp. 203–212. Springer, Cham (2024). https://doi.org/10.1007/978-3-031-61007-3_16

2. Conrardy, A., Capozucca, A., Cabot, J.: User modeling in model-driven engineering: a systematic literature review. J. Object Technol. **24**(2), 2:1–14 (2025). https://doi.org/10.5381/jot.2025.24.2.a12

3. Cummings, J.A., Sanders, L.: Introduction to Psychology. University of Saskatchewan Open Press (2019)

4. Feher, A., Vernon, P.A.: Looking beyond the big five: a selective review of alternatives to the big five model of personality. Pers. Individ. Differ. **169**, 110002 (2021). https://doi.org/10.1016/j.paid.2020.110002

5. Forbes, D.L.: Toward a unified model of human motivation. Rev. Gen. Psychol. **15**(2), 85–98 (2011). https://doi.org/10.1037/a0023483

6. Gaspar, A., Gil, M., Panach, I., Romero, V.: Towards a general user model to develop intelligent user interfaces. Multimedia Tools Appl. **83**, 1–34 (2024). https://doi.org/10.1007/s11042-024-18240-w

7. Grundy., J.C.: Impact of end user human aspects on software engineering. In: Proceedings of the 16th International Conference on Evaluation of Novel Approaches to Software Engineering - ENASE, pp. 9–20. SciTePress (2021). https://doi.org/10.5220/0010531800090020

8. Heckmann, D.: Ubiquitous User Modeling. IOS Press (2006)

9. Heckmann, D., Schwartz, T., Brandherm, B., Schmitz, M., Wilamowitz-Moellendorff, M.: Gumo – the general user model ontology. In: User Modeling, pp. 428–432 (2005). https://doi.org/10.1007/11527886_58

10. Hofstede, G.: Dimensionalizing cultures: the Hofstede model in context. Int. J. Behav. Med. **2** (2007). https://doi.org/10.9707/2307-0919.1014

11. Jiang, Y., Guo, Z., Rezazadegan Tavakoli, H., Leiva, L.A., Oulasvirta, A.: EyeFormer: predicting personalized scanpaths with transformer-guided reinforcement learning. In: Proceedings of the 37th Annual ACM Symposium on User Interface Software and Technology. Association for Computing Machinery (2024). https://doi.org/10.1145/3654777.3676436

12. Kaklanis, N., et al.: Towards standardisation of user models for simulation and adaptation purposes. Univers. Access Inf. Soc. **15**(1), 21–48 (2016). https://doi.org/10.1007/s10209-014-0371-2

13. Karam, R., Fraternali, P., Bozzon, A., Galli, L.: Modeling end-users as contributors in human computation applications. In: Abelló, A., Bellatreche, L., Benatallah, B. (eds.) MEDI 2012. LNCS, vol. 7602, pp. 3–15. Springer, Heidelberg (2012). https://doi.org/10.1007/978-3-642-33609-6_3

14. Lane, A., Beedie, C., Terry, P.: Distinctions between emotion and mood. Cogn. Emot. **19** (2005). https://doi.org/10.1080/02699930541000057

15. Liebel, G., Klünder, J., Hebig, R., Lazik, C., Nunes, I., Graßl, I., et al.: Human factors in model-driven engineering: future research goals and initiatives for MDE. Softw. Syst. Model. **23**(4), 801–819 (2024). https://doi.org/10.1007/s10270-024-01188-8

16. Liu, S., et al.: Applying a smartwatch to predict work-related fatigue for emergency healthcare professionals: machine learning method. West. J. Emerg. Med. Integr. Emerg. Care Popul. Health **24**(4) (2023). https://doi.org/10.5811/westjem.58139

17. Lochner, K., Eid, M.: Successful Emotions: How Emotions Drive Cognitive Performance. Springer, Wiesbaden (2016). https://doi.org/10.1007/978-3-658-12231-7

18. Marcus, A.: User interface design and culture. In: Usability and Internationalization of Information Technology, vol. 3, pp. 51–78 (2005). https://doi.org/10.1145/1556262.1556264

19. Mior, M.J.: Large language models for JSON schema discovery (2024). https://doi.org/10.48550/arXiv.2407.03286

20. Murgia, E., Pera, M.S., Landoni, M., Huibers, T.: Children on ChatGPT readability in an educational context: myth or opportunity? In: Adjunct Proceedings of the 31st ACM Conference on User Modeling, Adaptation and Personalization, pp. 311–316. Association for Computing Machinery (2023). https://doi.org/10.1145/3563359.3596996

21. Park, J.S., et al.: Generative agent simulations of 1,000 people (2024). https://doi.org/10.48550/arXiv.2411.10109

22. Patterson, R.D., Nimmo-Smith, I., Weber, D.L., Milroy, R.: The deterioration of hearing with age: Frequency selectivity, the critical ratio, the audiogram, and speech threshold. J. Acoust. Soc. Am. **72**(6), 1788–1803 (1982). https://doi.org/10.1121/1.388652

23. Planas, E., Daniel, G., Brambilla, M., Cabot, J.: Towards a model-driven approach for multiexperience AI-based user interfaces. Softw. Syst. Model. **20**(4), 997–1009 (2021). https://doi.org/10.1007/s10270-021-00904-y

24. Plocher, T., Rau, P.L.P., Choong, Y.Y., Guo, Z.: Cross-cultural design, Chap. 10, pp. 252–279. Wiley (2021). https://doi.org/10.1002/9781119636113.ch10

25. Purificato, E., Boratto, L., William De Luca, E.: User modeling and user profiling: a comprehensive survey (2024). https://doi.org/10.48550/arXiv.2402.09660

26. Robinson, L., et al.: Digital inequalities 3.0: emergent inequalities in the information age. First Monday (2020). https://doi.org/10.5210/fm.v25i7.10844

27. Svennberg, K., Ekman, J.: Structuring semi-structured data from building inspection reports using a large language model. In: Multiphysics and Multiscale Building Physics, pp. 508–513. Springer, Singapore (2025). https://doi.org/10.1007/978-981-97-8313-7_70

28. White, C., et al.: LiveBench: a challenging, contamination-free LLM benchmark (2024). https://doi.org/10.48550/arXiv.2406.19314

29. Yigitbas, E., Jovanovikj, I., Biermeier, K., Sauer, S., Engels, G.: Integrated model-driven development of self-adaptive user interfaces. Softw. Syst. Model. **19**(5), 1057–1081 (2020). https://doi.org/10.1007/s10270-020-00777-7

On the Use of LLMs to Explain Model Checking Counterexamples

Ezequiel José Veloso Ferreira Moreira[ID] and José Creissac Campos[✉][ID]

Departamento de Informática, Universidade do Minho & HASLab/INESC TEC, Braga, Portugal
id9620@alunos.uminho.pt, jose.campos@di.uminho.pt

Abstract. Formal verification has the potential to play a central role in the development of interactive safety-critical systems by providing rigorous guarantees about system behavior. However, the effective use of verification results remains a challenge in practice, particularly when these results must be interpreted by designers and domain experts who will not be formal methods specialists. This paper explores the use of Large Language Models (LLMs) to generate natural language explanations of counterexamples produced by model checking. More specifically, we present a study evaluating how different LLMs handle counterexamples produced from a range of formal models. Our focus is on the potential of LLMs to serve as mediators between formal verification tools and the multidisciplinary teams that design, develop, and validate interactive systems. The goal is to bridge the gap between formal outputs and human understanding. By examining the limitations and opportunities of the use of LLMs, we contribute to a broader discussion on integrating AI technologies into the engineering of trustworthy interactive systems.

Keywords: Safety-critical interactive systems · Model checking · Large Language Models (LLMs)

1 Introduction

Human-centered design (HCD) methods are essential for the development of interactive systems. However, as Thimbleby argues [40], in safety-critical domains, where failures can have severe consequences, HCD does not offer the level of rigor, repeatability, and analytical depth required to ensure system correctness and safety. HCD places emphasis on usability, user satisfaction, and

E.J.V.F. Moreira—Ezequiel Moreira acknowledges financial support from FCT – Fundação para a Ciência e Tecnologia, I.P., under grant 2023.01639.BD, funded by the ESF (European Social Fund) and by national funds through the Demography, Qualifications and Inclusion Program (PDQI).
J.C. Campos—Acknowledges financial support from FCT – Fundação para a Ciência e Tecnologia, I.P., under grant FCT/Mobility/1340210111/2024-25, funded by the European Union – NextGenerationEU.

C. Fayollas et al. (Eds.): EICS 2025, LNCS 16511, pp. 86–104, 2026.
https://doi.org/10.1007/978-3-032-26051-2_8

iterative refinement through empirical and qualitative methods such as proto-typing and user testing, rather than on ensuring formal correctness or eliminating all potential failure modes. While effective for creating intuitive and user-friendly systems, these methods were not designed to demonstrate that the system will behave safely and predictably under all conditions.

To meet these demands, the application of formal verification techniques to interactive systems has been proposed [5, 20, 43]. These techniques enable precise specification and exhaustive analysis of system behavior. They are not meant to replace human-centered methods but rather to complement them. The challenge lies in effectively integrating HCD with formal approaches in a manner that harnesses the strengths of both approaches.

One key obstacle to this integration is communication. The outputs of formal verification are typically expressed in formal notations that require specialized expertise to be interpreted. While this is the natural consequence of the need for formality and rigor (cf. [13]), it creates a barrier for domain experts and design practitioners who may lack formal methods training and is one of the reasons for the practical challenges non-expert and domain engineers face when interpreting formal verification results [27].

We are particularly interested in model checking approaches due to their automated nature. Once the model and the requirements are expressed in adequate formal languages, the verification is performed without human intervention. Model checkers produce counterexamples when verification fails. These counterexamples illustrate system behaviors that falsify the property being proved, and can be interpreted as illustrating scenarios where the property being verified does not hold. These counterexamples then need to be analyzed and understood to determine the cause of failure.

In previous work [35], we have proposed an approach to generating natural language explanations of model checking counterexamples, and reported on some initial experimentation with LLMs. In this paper, we explore further how LLMs can support this process. We investigate the use of several LLMs to translate model checking counterexamples into natural language descriptions that are more accessible to non-expert users. Our focus is on evaluating the correctness of the generated natural language descriptions, thus contributing to a broader discussion on the feasibility of integrating AI technologies into the engineering of trustworthy interactive systems.

2   Related Work

Before proceeding to the evaluation of LLMs, this section provides a review of work on generating explanations for model checking counterexamples.

2.1   Model Checking

Model checking [10] is an automated verification technique that systematically explores the state space of a formal model of a system (or, in the case of software model checkers [9,22], the state space derived directly from source code) to determine whether it satisfies a formal specification. These specifications are typically expressed in temporal logic such as Computational Tree Logic (CTL) [8] or Linear Time Logic (LTL) [17]. When the specification is not satisfied, the model checker attempts to produce a counterexample, which is an execution trace demonstrating how the model's behavior leads to the violation of the property.

Model checking has been successfully applied to interactive systems, particularly in domains where correctness and safety are critical, such as avionics (e.g., airplane cockpits [12]), aerospace (e.g., ground segments of satellite launcher systems [7]), healthcare (e.g., medical devices [31]), and automotive (e.g., a cruise control system [3]). A recurring concern, however, is the need to make the formal approaches more accessible to designers and domain experts. This needs to be done at three levels: developing the model, expressing properties for verification, and interpreting verification results. In this paper, we focus on the latter. That is, in supporting the interpretation of counterexamples.

2.2   Model Checking Counterexamples

Model checking counterexamples must be interpreted to identify the problem they illustrate. This can be a problem with the model (the system's design) or with the property (either with the requirement itself or how it was encoded in the logic).
However, interpreting counterexamples can be challenging:

- Execution traces are typically presented as sequences of abstract states and transitions, and must be related back to real-world behavior.
- The use of temporal logics adds further complexity, requiring users to understand the formal semantics of the temporal operators to interpret why a property fails.
- The traces often do not explain causality, leaving users to infer the root issue from raw data.
- In large models, the volume of states and transitions can be overwhelming, and tool limitations (such as linear trace outputs for branching properties) further obscure understanding.
- False negatives, due to abstraction errors or trade-offs, incomplete state space exploration, or assumptions used, often aggravate the problem.

These challenges are particularly relevant for domain experts and other stakeholders who are not familiar with formal methods, but are essential for assessing the practical significance of the verification results.

2.3   Explaining Model Checking Counterexamples

A variety of approaches have been proposed to make model checking counterexamples more understandable (for a review, see Kaleeswaran et al. [26]). Tools such as STATEMATE [19] and UPPAAL2k [1] represent models using

Statecharts. The IVY workbench [6] adds support for state transition tables and UML [16] activity diagrams. Work on minimizing and slicing counterexamples [25] reduces the trace size, allowing engineers to focus on the relevant portion of an execution. However, the trace is still presented at the model level.

Other efforts aim to make counterexamples more accessible by presenting them in formats that are closer to the problem domain. Proposals include the use of timing diagrams [24,36,37], function block diagrams [36,37], or task models [2]. In many cases, these representations can be explored by manipulating the representation (for example, filtering or highlighting specific parts). This helps users, especially those unfamiliar with formal methods, understand the behavior illustrated by the counterexample to identify where and how violations occur.

Explanation-enhancing techniques augment counterexamples with causal and semantic information, further helping users understand why a property fails. Tools like `explain` [18] can determine causal dependencies between predicates in an execution by identifying the subformulas whose false values are necessary and sufficient to cause the overall formula to fail. `explain` complements this with the capability to generate valid executions that are as similar as possible to the counterexample, and new counterexamples that are as different as possible from the original one.

Beyond visualizations, there is work on automatically generating structured natural-language explanations of counterexamples from model-checker artifacts. The typical approach is to define domain-specific controlled natural languages in which the counterexamples can be expressed. This type of approach has been applied to several domains. For example, railway systems [30,41], normative documents [4] and robotics [15]. A different approach was proposed by Moreira and Campos, which resorts to generic property specification patterns to support explanation generation [33,34].

Natural language explanations offer a promising solution by translating formal traces into accessible narratives, helping users better understand the underlying issues. One question is what role Artificial Intelligence (AI) techniques might play in generating natural language explanations. In the past, the need to train models meant that developing adequate AI models could be too costly or even unfeasible, due to a lack of enough training data. Current developments on LLMs, however, raise the question of whether their capabilities to generate human-level text can be leveraged to create such explanations.

Indeed, the advent of large pretrained language models has stimulated a new wave of research applying LLMs to formal-methods tasks. Work ranges from using LLMs to translate natural language requirements into temporal logics [11, 29,32], to LLM-assisted invariant/lemma generation and theorem proving [28, 38,44]. However, to the best of our knowledge, the use of LLMs to explain the results of formal verification has not yet been addressed.

3   Background

As explained above, this paper investigates the feasibility of applying LLMs to generate explanations for counterexamples. This section provides relevant background information that will be useful in understanding the evaluation that was carried out.

3.1   Property Specification Patterns

Property specification patterns, initially proposed by Dwyer et al. [14], provide a systematic framework for expressing common behavioral properties of software and hardware systems in a reusable and formalized manner. These patterns abstract temporal logic formulas into high-level templates (e.g., *absence*, *existence*, *universality*, and *response*), facilitating the specification and verification of critical system requirements. For each pattern, besides its description and intent, one or more parameterized formulas in temporal logic (e.g., LTL, CTL) are provided. While the original patterns targeted finite-state systems, subsequent work by Harrison et al. [21] proposed additional patterns tailored to interactive systems, addressing properties such as *undo* and *feedback*.

3.2   An Approach to Counterexample Explanation

In previous work [33,35] we propose a pattern-based approach to generating natural language explanations from model checking counterexamples produced by the NuSMV model-checker (see Fig. 1). The approach was designed to support the IVY workbench tool, which uses NuSMV as its verification engine. The tool uses MAL interactors as its modeling language. Herein, it is enough to understand that a MAL interactor has a state defined as a set of typed attributes, actions that act on that state, and a set of (modal) axioms that express how the actions change the interactor's state.

The approach makes use of a library of property specification patterns to identify the type of property that generated the counterexample. Using a query language to obtain information from the trace (cf. [34]), plus domain information obtained from the model (through the use of designations – cf. [23]), a natural language explanation is generated.

Two aspects of the patterns-based approach are relevant in this context. The first aspect is that the structure of the counterexamples produced by model checkers will have specific characteristics that can be deduced from the structure of the property that originated it (for example, the position of relevant states in the trace). These characteristics can be used to narrow down the scope of the search for relevant information in the counterexample to a much smaller set of states that are relevant to the problem.

The second aspect is that domain-specific information can significantly improve the quality of explanations for counterexamples. This is due to the above-discussed need for specialized expertise to interpret the formal notations used. Associating domain knowledge with relevant parts of the formal language

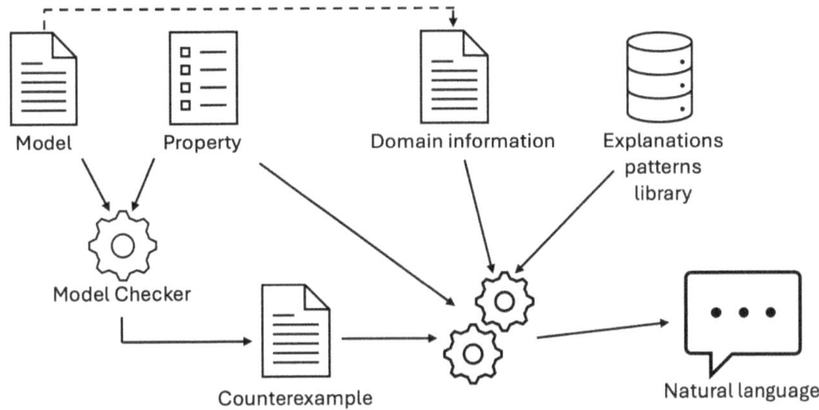

Fig. 1. The process of generating natural language explanations (from [33])

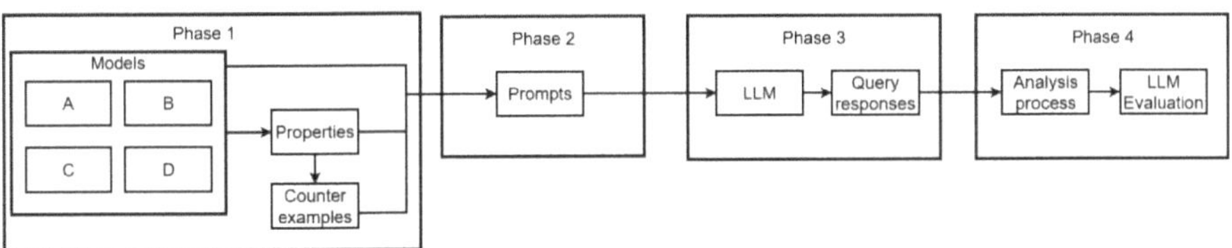

Fig. 2. LLM evaluation process

enables a clearer understanding of what they represent, facilitating the creation of more comprehensible explanations.

To allow for a later comparison of the results of this analysis with those of Moreira and Campos, we will employ a comparable approach, based on a MAL model, property templates, and domain information, to generate the prompts that will be used to test the LLMs.

3.3 Large Language Models

LLMs are a type of AI model based on the Transformer architecture [42], and trained on large unlabeled text corpora through generative pre-training [39]. This approach enables them to capture complex linguistic patterns and both understand and generate coherent, contextually relevant natural language.

LLMs have achieved widespread adoption across industries and research, powering platforms such as ChatGPT, Claude, and Gemini. Their versatility spans applications from conversational agents and code generation to content creation and knowledge synthesis.

4   Approach to LLMs Evaluation

The method created to evaluate how LLMs transform counterexamples into natural language descriptions consists of four phases, summarized in Fig. 2:

1. Selecting the formal models and properties to generate the counterexamples to be used in the evaluation.
2. Creating a series of prompts for the LLMs.
3. Submitting the created prompts to the LLMs.
4. Evaluating the responses.

4.1   Formal Models and Properties

The first phase is selecting the formal models and properties that will be used in the evaluation. These formal models represent interactive systems with a growing level of complexity, and over each of these, a series of different properties was verified. These properties were all derived from a series of common property patterns, and all of them are violated by the selected models in different ways, thus producing a set of counterexamples that need to be explained.

Four models of increasing complexity were defined and verified: the model of a remote controller for a garage door (Model A), a model capturing the management of window visibility in an email client (Model B), the model of a multimeter's user interface (Model C), and the model of a washing machine's user interface (Model D).

Model A featured two attributes, four actions, and twelve axioms. One property was defined and verified. Model B featured two interactors, one with three attributes, six actions, and twelve axioms, and another aggregating two instances of the former, with just one additional axiom. Three properties were defined and verified. Model C featured six attributes, nine actions, and seventeen axioms. Six properties were defined and verified. Finally, Model D features nine attributes, forty-four actions, and ninety axioms. A total of thirty properties were defined and verified on this model. All properties were false, so model checking of Model A generated one property/counterexample pair, Model B three pairs, Model C six pairs, and Model D thirty pairs.

4.2   Creating the Prompts

With this data, the second phase consists of creating a series of prompts for the LLMs. Four different types of prompts were created, each with distinct characteristics that enable evaluation of the LLMs' performance under various conditions.

– The first prompt type (PROMPT1) asks the LLM to explain the counterexample and determine the original problem with the model by first providing several pieces of information: the formal model; domain information explaining the meaning of the types, attributes and actions in the model; the properties paired with their respective counterexamples; and, finally, an explanation of where in the counterexamples relevant information can be found, depending on the pattern matching the property. Details on the specific format used can be found in the Appendix.

- The second prompt type (PROMPT2) is identical to the first, except that it does not include the formal model. The goal was to investigate whether the presence of the model would improve the quality of the output.
- The third prompt type (PROMPT3) contains the same information as the second prompt, but presented differently. While prompts of type PROMPT1 and PROMPT2 are expressed in a natural language style, prompts of type PROMPT3 adopt a more structured style based on tagging relevant components of the prompt. The goal was to investigate whether this could have any impact on the results.
- The fourth and final prompt type (PROMPT4) is composed of a prelude prompt and a data prompt. The prelude prompt contains information that explains, in detail, how to read each section of the data prompt and where in the counterexample different types of properties have the relevant data expressed. The data prompt is split into three sections, containing information about the domain, the model, the properties, and their respective counterexamples. This final prompt type extends the previous ones by additionally asking the LLM to propose a valid solution for the problem, alongside explaining the counterexample and determining the problem with the model.

Each prompt contained all pairs for the given model.

4.3   Submitting the Prompts

The third phase consisted of submitting the created prompts to the selected LLMs. The LLMs would then respond to the prompts, and these responses were stored for analysis and evaluation.

For prompts types PROMPT1 to PROMPT3, models A, C, and D were used. Each prompt was submitted ten times to each LLM. Hence, for example, for Model D, for which we had 30 properties, a total of 300 property/counterexample pairs was submitted using each of the prompt types PROMPT1 to PROMPT3. However, due to limitations in the size of the messages processed by the LLMs, it was only possible to obtain responses for a subset of the pairs for this model: 83, 89 and 86 pairs for prompt types PROMPT1, PROMPT2 and PROMPT3 from ChatGPT; and 43, 45 and 70 pairs from Claude, respectively.

To address this issue, fewer properties were used for the last prompt type (PROMPT4) and the input was split into prelude and data, as explained above. For this prompt type, Model B replaced the simplest model of the previous three prompt types (Model A). This was done because Model A was found to be so simple that it allowed only the creation of the most trivial properties to be verified. This meant that most property/counterexample pairs obtained from this model were too simple to allow for relevant information to be obtained from the prompts.

For the first model, three property/counterexample pairs were defined using each of the prompt types. For the second model, six property/counterexample pairs were defined using each of the prompt types, as before. For the third model, five property/counterexample pairs were defined using each of the prompt types. As before, each was repeated ten times.

The answers given by each LLM were gathered for analysis in the next stage.

4.4 Evaluating the Responses

The fourth phase consisted of evaluating the responses produced in the previous phase. This evaluation focused on determining the LLM's ability to explain the provided counterexamples and identify what problems led to their creation. These criteria were chosen as a means to evaluate different aspects of the process of analyzing counterexamples and, from that analysis, refine the formal model.

- The explanations produced by the LLMs allow us to assess their ability to interpret the data.
- The identification of the underlying problem allows us to evaluate their ability to reason with formal models and accurately determine where and how the problem occurred.

Additionally, the fourth prompt type was tested for its ability to propose valid solutions to the problems it identified, with the goal of assessing whether the LLM could draw its own conclusions and apply them to solve the errors that led to the counterexamples.

Using the answers gathered in the previous phase, the performance of each selected LLM was evaluated across the different prompt types and tested models. The evaluation was conducted by the first author and subsequently discussed and validated with the second author.

5 Selected LLMs for Evaluation

With the evaluation process determined, four LLM systems were selected as good candidates for evaluation: Google Gemini 1.5 Flash, ChatGPT 4o, Claude 3.5 Sonnet, and LlamaFile with LLaVa-v1.5-7b-q4. The first three candidates were chosen due to their widespread usage, while the last candidate was chosen to test local LLMs (i.e., LLMs that run directly on the user's machine rather than on an external server).

Before proceeding with the full evaluation, a pilot testing phase was conducted, consisting of the first three types of queries applied to the simplest model (Model A). The goal was to determine whether the selected systems could process this simplest model and the smallest set of tests. Any LLM failing to analyze this case was excluded from further evaluation. The simplicity of this test implies that such a failure indicates an inability to handle more realistic formal models, which are much larger and more complex than this trivial case, rendering the system unsuitable for subsequent testing.

Table 1. Error rate by LLM

LLM	Explanation	Problem identification
ChatGPT	10%	24%
Claude	11%	33%

From these initial tests, only ChatGPT 4o and Claude 3.5 Sonnet were selected for further exploration. Google Gemini was found to produce explanations of substantially lower quality than the selected LLMs, with considerably more hallucinations. As for LLaVa-v1.5-7b-q4, the LLM was far too slow to run on the hardware available for the test procedure (Intel Core i5-8250U with 12 GB of RAM). As such, it was also deemed unfit for the test, not due to the quality of the explanations produced, but due to its practical applicability under the testing conditions.

With the initial testing concluded, we could proceed with the evaluation of the selected LLM systems.

6   Evaluation Results

Tables 1, 2 and 3 present some initial results of the analysis of the LLMs' outputs. Two main aspects are presented in the tables: the error rates for explaining the counterexample (first column) and the error rates for problem identification (second column).

It should be noted that, for prompts of the first three types, "Problem identification" refers to the LLM's ability to determine what the problem with the model was. In contrast, the fourth type includes this, as well as the proposed solutions the LLM gave to the problems identified in the model. This merging was done because in every case where the LLM incorrectly interpreted the model, there was also an incorrect proposed solution; and every case where an acceptable interpretation of the model was presented was accompanied by an acceptable proposed solution.

As shown in Table 1, the performance of both models is similar, with the results being considerably worse when it comes to identifying problems. The percentage of erroneous explanations of the counterexamples, although considerably smaller (10%–11%), is still substantial, particularly if we consider a safety-critical context.

Table 2 shows the results by model. As can be seen, model C has considerably worse results, even though we considered it less complex than model D. The reasons for this need to be investigated, but they may be related to the fact that model D represents a washing machine, about which the LLM may have more knowledge than it does about multimeters (or the knowledge might be more aligned with the model). Again, except for model A, which was trivial, the number of erroneous explanations of the trace is non-negligible.

Table 2. Error rate by model

Model	Explanation	Problem identification
A (garage door controller)	0%	8%
B (email client windows)	5%	68%
C (multimeter)	30%	37%
D (washing machine)	6%	20%

Table 3. Error rates by prompt type

prompt type	Explanation	Problem identification
PROMPT1	13%	16%
PROMPT2	23%	25%
PROMPT3	12%	39%
PROMPT4	17%	59%

Finally, Table 3 shows the results by prompt type. It can be observed that including the model in the prompt (PROMPT1) has a positive impact on the results. Still, its absence can be compensated for, at least regarding the explanation of the counterexamples, by providing more structure to the prompt.

While not evident on the tables, analysis of the collected data showed that Claude 3.5 Sonnet produced better explanations in the simpler models and had consistent performance across every prompt type. At the same time, ChatGPT performed much more consistently across all models and showed a significant improvement in response quality for the fourth type of prompt.

Each LLM produced explanations that were, in most cases, easy to understand and interpret, as well as a good analysis of the presented counterexamples and underlying problems that led to their creation. For example, one LLM was able to identify a situation where a property expressed over the washing machine model:

$$AG(state \neq errorState)$$

expressed an unrealistic requirement (that the machine would never go into an error state). The LLM identifies the problem as:

> *"The property requires that the machine never enters an error state, which is unrealistic as error states are necessary to handle exceptional conditions like excessive loads."*

and proposed that no corrective action be taken:

> *"This appears to be an intentionally incorrect test property. Error states are a crucial part of the washing machine's safety and operational design. The property should not be modified – it's correctly failing as error states are necessary."*

However, both LLMs displayed a non-negligible number of erroneous analyses of counterexamples and underlying problems, with even more serious issues in their proposed solutions. For example, during the analysis of a counterexample for the property:

$$AG(state \neq errorState \implies error = 0)$$

(if the machine is not in an error state, the *error* attribute must be zero), the LLM concludes that:

> *"The problem here is that 'error' is updated before the transition into 'errorState', violating the property."*

In reality, the error was that for one of the possible error states, the value of the attribute *error* was not being properly updated (leading to it exiting the error state without changing the attribute to 0). Hence, the property was violated because the attribute was not being updated on leaving an error state, not because it was being updated too early when entering one.

Additionally, a substantial amount of effort was required to detect errors made by the LLM in the evaluation process, as the errors became more subtle in more complex models. Furthermore, a notable increase in the subtlety and complexity of the error detection process was observed when comparing the analysis of the fourth prompt with that of the preceding prompts, likely due to the increased context provided to the LLMs.

An example of such an error occurred when analyzing whether the machine correctly progresses through its programs. To test that the duration of the program decreases, we attempted to prove:

$$AG(duration = medium \implies AX(duration = short))$$

which failed with a counterexample showing that the machine could go into an error state. The LLM's analysis of the counterexample determined that:

> *"The property fails because a **rinsing error** interrupts the intended duration progression. This is a valid situation and accounted for in the model via: per(rinsingErrorMedium) $\implies$ state=rinsingPhase & duration=medium"*

which is a valid conclusion. The issue with the LLM's answer is the property it suggested as the solution to the problem: that the tested property should be changed to

$$AG\big((duration = medium \ \& \ state \neq errorState) \implies$$
$$AX(duration = short \mid state = errorState)\big)$$

(i.e., either the duration decreases or the machine went into an error state).

At first inspection, the proposed property constitutes a valid solution to the problem. Nevertheless, it fails to preserve the original intent, namely, to ensure that a state with *duration = medium* is subsequently followed by a state with *duration = short*. In contrast, the proposed property is satisfied by the model's current behavior due to the presence of the error case. Yet, it does not guarantee the desired behavior in the absence of such errors. Consequently, this property represents an incorrect formulation of the intended requirement.

These factors strongly suggest that while the fourth prompt may have improved the overall results, it also made the errors that were committed harder to detect. They also suggest that increasing the complexity of the model leads to a higher error rate and that these errors become more difficult to detect.

7   Threats to Validity

The study highlights both the potential and the limitations of LLMs in generating natural-language explanations of counterexamples. However, several factors should be considered when interpreting the results, as they may influence the generalizability and applicability of the findings.

7.1   LLM Selection

The evaluation considered a small number of LLMs, which, while representative of current state-of-the-art systems, limits the generalizability of the results. LLMs are continuously evolving, and their performance may change as models are updated or retrained. However, their inherent stochastic nature remains unchanged, meaning that variability in responses is an intrinsic property independent of model version, which partially preserves the relevance of the evaluation. The selected LLMs provide meaningful insights into the studied scenarios, although the results should be interpreted with this context in mind.

7.2   Model and Prompt Characteristics

The formal models used vary in size and complexity but remain relatively small and straightforward. Additionally, the prompts were not systematically varied in length or complexity. As a result, the performance observed in this study may differ for larger and more complex industrial models. Nonetheless, these models and prompts are sufficient to demonstrate the current limitations of LLMs when applied to formal-model analysis tasks.

7.3   Domain-Specific Training and Knowledge Bias

The LLMs employed were general-purpose systems and were not specifically trained on the domain under study. No additional training was performed beyond the data segmentation applied to the fourth prompt type. While this may influence absolute performance, the evaluation still illustrates the capability of current LLMs to reason over formal models without specialized adaptation.

Furthermore, the LLMs' large-scale pre-training on extensive, general-domain data may have a dual effect: on the one hand, it can support a more accurate interpretation of the formal models due to prior exposure to information about the domain; on the other hand, it may introduce unintended biases, leading the LLM to assume behaviors or semantics that differ from those explicitly defined in the formal model.

7.4   Validation of Results

The results were not independently validated by external researchers or domain experts. Consequently, potential biases in interpretation or unnoticed errors may remain. However, the evaluation methodology is clearly defined and reproducible, enabling future studies to extend or confirm the findings.

8   Conclusion

Formal verification, and in particular model checking, provides rigorous guarantees of system correctness, producing counterexamples when properties fail. Interpreting these counterexamples is a critical step in understanding system behavior and addressing potential errors, but can often be challenging for multidisciplinary teams without expertise in formal methods. This study investigated whether LLMs could assist by generating natural language explanations of counterexamples to models of interactive systems, potentially making them more accessible to multidisciplinary teams and supporting integration with human-centered design processes.

The evaluation shows that, while in most cases LLM-generated explanations were understandable, a significant proportion contained errors, including misinterpretations of the counterexamples and incorrect identification of the underlying issues. Given that model checking is employed to provide formal guarantees of correctness, especially in safety-critical contexts, relying on LLMs in this role is currently unsafe. Erroneous explanations could mislead designers and engineers, undermining the formal guarantees that these methods are meant to provide.

These results indicate that LLMs cannot yet be relied upon as intermediaries between formal verification tools and the multidisciplinary teams involved in the design, development, and validation of interactive systems. Further research is required to determine whether the observed limitations are intrinsic to current LLM architectures or if they can be mitigated. Possible directions include: exploring domain-specific fine-tuning of LLMs, developing hybrid workflows in which LLM-generated explanations are validated or corrected by automated reasoning tools or experts, and evaluating the practical impact of LLM assistance in real-world verification and design scenarios.

In conclusion, while LLMs show potential to bridge gaps in accessibility, the current generation of models is not sufficiently reliable for critical verification tasks. Future work must focus on ensuring correctness and robustness before such tools can safely support the interpretation of formal verification outputs.

Appendix: Format of the Prompt1 prompt type

This appendix illustrates one prompt of type PROMPT1. For brevity's sake, parts of the prompt have been omitted or replaced with descriptive text between square brackets.

```
I would like you to produce some explanations for counterexamples
produced by the NuSMV model checker, so that I may understand
where the problem lies.

The model that was verified is depicted below, starting the line
below "START MODEL FILE" and ending the line before the word "END
MODEL FILE".

START MODEL FILE
[model file that was verified]
END MODEL FILE


[Very small explanation of what the model that was tested is in
natural language]

For this model, the types represent the following:
[Type identifier 1] - [natural language explanation of what the
type represents]
[Type identifier 2] - [natural language explanation of what the
type represents]
[...]

For this model, the attributes represent the following:
[Attribute identifier 1] - [natural language explanation of what
the attribute represents]
[Attribute identifier 2] - [natural language explanation of what
the attribute represents]
[...]

For this model, the actions represent the following:
[Action identifier 1] - [natural language explanation of what the
action represents]
[Action identifier 2] - [natural language explanation of what the
action represents]
[...]

This model has had the following CTL properties verified over it:
[list of the CTL properties]

For each CTL property shown above, the counterexamples that were
produced by NuSMV are depicted below sequentially, between the
lines that have the words "START COUNTEREXAMPLES" and "END
COUNTEREXAMPLES".
```

```
START COUNTEREXAMPLES
[list of counterexamples, ordered in the same order as the
properties above]
END COUNTEREXAMPLES

It should be noted that the counterexamples shown have been
simplified to avoid repetition, such that if the value of a
property does not change between states, then it is not written
in the counterexample.

Additionally, the counterexamples produced by each property have
the following characteristics:
- Properties of type AG([X]) have [X] not being verified only on
the last state of the counterexample.
[...]

With all of this, can you please explain the counterexample?
Additionally, can you determine what the problem with the
original model was?
```

References

1. Amnell, T., et al.: UPPAAL - now, next, and future. In: Cassez, F., Jard, C., Rozoy, B., Ryan, M.D. (eds.) MOVEP 2000. LNCS, vol. 2067, pp. 99–124. Springer, Heidelberg (2001). https://doi.org/10.1007/3-540-45510-8_4
2. Bolton, M.L., Bass, E.J.: Using task analytic models to visualize model checker counterexamples. In: 2010 IEEE International Conference on Systems, Man and Cybernetics, pp. 2069–2074. IEEE (2010). https://doi.org/10.1109/ICSMC.2010.5641711
3. Bolton, M., Siminiceanu, R., Bass, E.: A systematic approach to model checking human–automation interaction using task analytic models. IEEE Trans. Syst. Man Cybern. Part A Syst. Hum. **41**(5) (2011). https://doi.org/10.1109/TSMCA.2011.2109709
4. Camilleri, J.J., Haghshenas, M.R., Schneider, G.: A web-based tool for analysing normative documents in English. In: Proceedings of the 33rd Annual ACM Symposium on Applied Computing, pp. 1865–1872. ACM (2018). https://doi.org/10.1145/3167132.3167331
5. Campos, J.C., Harrison, M.D.: Formal approaches for interactive systems. In: Vanderdonckt, J., Palanque, P., Winckler, M. (eds.) Handbook of Human Computer Interaction. Springer, Cham (2025). https://doi.org/10.1007/978-3-319-27648-9_120-1
6. Campos, J., Harrison, M.D.: Interaction engineering using the IVY tool. In: ACM Symposium on Engineering Interactive Computing Systems (EICS 2009), pp. 35–44. ACM (2009). https://doi.org/10.1145/1570433.1570442
7. Campos, J., Sousa, M., Alves, M., Harrison, M.: Formal verification of a space system's user interface with the IVY workbench. IEEE Trans. Hum. Mach. Syst. **46**(2), 303–316 (2016). https://doi.org/10.1109/THMS.2015.2421511

8. Clarke, E.M., Emerson, E.A., Sistla, A.P.: Automatic verification of finite-state concurrent systems using temporal logic specifications. ACM Trans. Program. Lang. Syst. **8**(2), 244–263 (1986). https://doi.org/10.1145/5397.5399

9. Clarke, E., Kroening, D., Lerda, F.: A tool for checking ANSI-C programs. In: Jensen, K., Podelski, A. (eds.) Tools and Algorithms for the Construction and Analysis of Systems, pp. 168–176. Springer, Heidelberg (2004). https://doi.org/10.1007/978-3-540-24730-2_15

10. Clarke Jr., E.M., Grumberg, O., Peled, D.A.: Model Checking. MIT Press (1999)

11. Cosler, M., Hahn, C., Mendoza, D., Schmitt, F., Trippel, C.: nl2spec: Interactively translating unstructured natural language to temporal logics with large language models. In: Enea, C., Lal, A. (eds.) Computer Aided Verification, pp. 383–396. Springer, Cham (2023). https://doi.org/10.1007/978-3-031-37703-7_18

12. Degani, A., Heymann, M.: Formal verification of human-automation interaction. Hum. Fact. **44**(1), 28–43 (2002). https://doi.org/10.1518/0018720024494838

13. Dijkstra, E.W.: On the foolishness of "natural language programming". In: Bauer, F.L., et al. (eds.) Program Construction. LNCS, vol. 69, pp. 51–53. Springer, Heidelberg (1979). https://doi.org/10.1007/BFb0014656

14. Dwyer, M.B., Avrunin, G.S., Corbett, J.C.: Patterns in property specifications for finite-state verification. In: Proceedings of the 21st International Conference on Software Engineering, ICSE '99, pp. 411–420. ACM (1999). https://doi.org/10.1145/302405.302672

15. Feng, L., Ghasemi, M., Chang, K.W., Topcu, U.: Counterexamples for robotic planning explained in structured language. In: 2018 IEEE International Conference on Robotics and Automation (ICRA), pp. 7292–7297 (2018). https://doi.org/10.1109/ICRA.2018.8460945

16. Fowler, M.: UML Distilled, 3rd edn. The Addison-Wesley Object Technology Series. Addison-Wesley (2003)

17. Gabbay, D., Pnueli, A., Shelah, S., Stavi, J.: On the temporal analysis of fairness. In: Proceedings of the 7th ACM SIGPLAN-SIGACT Symposium on Principles of Programming Languages, pp. 163–173. ACM (1980). https://doi.org/10.1145/567446.567462

18. Groce, A., Kroening, D., Lerda, F.: Understanding counterexamples with `explain`. In: Alur, R., Peled, D.A. (eds.) CAV 2004. LNCS, vol. 3114, pp. 453–456. Springer, Heidelberg (2004). https://doi.org/10.1007/978-3-540-27813-9_35

19. Harel, D., et al.: STATEMATE: a working environment for the development of complex reactive systems. In: 11th International Conference on Software Engineering, pp. 396–406. IEEE (1988). https://doi.org/10.1109/ICSE.1988.93720

20. Harrison, M., Thimbleby, H. (eds.): Formal Methods in Human-Computer Interaction. Cambridge Series on Human-Computer Interaction. Cambridge University Press (1990)

21. Harrison, M., Masci, P., Campos, J.: Verification templates for the analysis of user interface software design. IEEE Trans. Softw. Eng. **45**(8), 802–822 (2019). https://doi.org/10.1109/TSE.2018.2804939

22. Havelund, K., Pressburger, T.: Model checking java programs using Java PathFinder. Int. J. Softw. Tools Technol. Transf. **2**(4), 366–381 (2000). https://doi.org/10.1007/s100090050043

23. Jackson, M., Zave, P.: Deriving specifications from requirements: an example. In: 1995 17th International Conference on Software Engineering, p. 15 (1995). https://doi.org/10.1145/225014.225016

24. Jee, E., Jeon, S., Cha, S., Koh, K., Yoo, J., Park, G., et al.: FBDverifier: interactive and visual analysis of counterexample in formal verification of function block diagram. J. Res. Pract. Inf. Technol. **42**(3), 171–188 (2010)
25. Jhala, R., Majumdar, R., Leino, K.R.M., Sankaranarayanan, S.: Path slicing: a dataflow technique for generating small, precise counterexamples. In: Proceedings of the 2005 ACM SIGPLAN Conference on Programming Language Design and Implementation (PLDI). ACM (2005). https://doi.org/10.1145/1064978.1065016
26. Kaleeswaran, A.P., Nordmann, A., Vogel, T., Grunske, L.: A systematic literature review on counterexample explanation. Inf. Softw. Technol. **145** (2022). https://doi.org/10.1016/j.infsof.2021.106800
27. Kaleeswaran, A.P., Nordmann, A., Vogel, T., Grunske, L.: A user study for evaluation of formal verification results and their explanation at Bosch. Empir. Softw. Eng. **28**(5), 125 (2023). https://doi.org/10.1007/s10664-023-10353-4
28. Li, Z., Sun, J., Murphy, L., Su, Q., Li, Z., Zhang, X., et al.: A survey on deep learning for theorem proving. In: First Conference on Language Modeling, COLM 2024 (2024). https://doi.org/10.48550/arXiv.2404.09939
29. Liu, J.X., et al.: Grounding complex natural language commands for temporal tasks in unseen environments. In: Conference on Robot Learning (CoRL) (2023). https://doi.org/10.48550/arXiv.2302.11649
30. Luteberget, B., Camilleri, J.J., Johansen, C., Schneider, G.: Participatory verification of railway infrastructure by representing regulations in RailCNL. In: Cimatti, A., Sirjani, M. (eds.) SEFM 2017. LNCS, vol. 10469, pp. 87–103. Springer, Cham (2017). https://doi.org/10.1007/978-3-319-66197-1_6
31. Masci, P., Ayoub, A., Curzon, P., Harrison, M.D., Lee, I., Thimbleby, H.W.: Verification of interactive software for medical devices: PCA infusion pumps and FDA regulation as an example. In: Forbrig, P., Dewan, P., Harrison, M., Luyten, K. (eds.) ACM SIGCHI Symposium on Engineering Interactive Computing Systems, EICS'13, pp. 81–90. ACM (2013). https://doi.org/10.1145/2494603.2480302
32. Mendoza, D., Hahn, C., Trippel, C.: Translating natural language to temporal logics with large language models and model checkers. In: 2024 Formal Methods in Computer-Aided Design (FMCAD), pp. 119–129. TU Wien Academic Press (2024). https://doi.org/10.34727/2024/isbn.978-3-85448-065-5_17
33. Moreira, E., Campos, J.: Explaining temporal logic model checking counterexamples through the use of structured natural language. In: Engineering Interactive Computer Systems. LNCS, vol. 14517, pp. 1–19. Springer, Cham (2024). https://doi.org/10.1007/978-3-031-59235-5_15
34. Moreira, E., Campos, J.: A language for explaining counterexamples. In: 13th Symposium on Languages, Applications and Technologies (SLATE 2024). OpenAccess Series in Informatics, vol. 120, pp. 11:1–11:14. Schloss Dagstuhl - LZI (2024). https://doi.org/10.4230/OASIcs.SLATE.2024.11
35. Moreira, E., Campos, J.: On the role of generative AI in explaining model checking counterexamples. In: Zaina, L., et al. (eds.) Engineering Interactive Computer Systems. LNCS, vol. 15518, pp. 138–158. Springer, Cham (2025). https://doi.org/10.1007/978-3-031-91760-8_10
36. Ovsiannikova, P., Buzhinsky, I., Pakonen, A., Vyatkin, V.: OERITTE: user-friendly counterexample explanation for model checking. IEEE Access **9**, 61383–61397 (2021). https://doi.org/10.1109/ACCESS.2021.3073459
37. Pakonen, A., Buzhinsky, I., Vyatkin, V.: Counterexample visualization and explanation for function block diagrams. In: 2018 IEEE 16th International Conference on Industrial Informatics (INDIN), pp. 747–753. IEEE (2018). https://doi.org/10.1109/INDIN.2018.8472025

38. Pirzada, M.A.A., Reger, G., Bhayat, A., Cordeiro, L.C.: LLM-generated invariants for bounded model checking without loop unrolling. In: Proceedings of the 39th IEEE/ACM International Conference on Automated Software Engineering, pp. 1395–1407. ACM (2024). https://doi.org/10.1145/3691620.3695512
39. Radford, A., Narasimhan, K., Salimans, T., Sutskever, I.: Improving language understanding by generative pre-training. Technical report, OpenAI (2018). arXiv:1801.06146
40. Thimbleby, H.: User-centered methods are insufficient for safety critical systems. In: Holzinger, A. (ed.) HCI and Usability for Medicine and Health Care. LNCS, vol. 4799. Springer, Heidelberg (2007). https://doi.org/10.1007/978-3-540-76805-0_1
41. van den Berg, L., Strooper, P., Johnston, W.: An automated approach for the interpretation of counter-examples. Electron. Notes Theor. Comput. Sci. **174**(4), 19–35 (2007). https://doi.org/10.1016/j.entcs.2006.12.027
42. Vaswani, A., Shazeer, N., Parmar, N., Uszkoreit, J., Jones, L., Gomez, A.N., et al.: Attention is all you need. Adv. Neural. Inf. Process. Syst. **30**, 6000–6010 (2017)
43. Weyers, B., Bowen, J., Dix, A., Palanque, P. (eds.): The Handbook of Formal Methods in Human-Computer Interaction. Springer, Human-Computer Interaction Series (2017)
44. Wu, G., Cao, W., Yao, Y., Wei, H., Chen, T., Ma, X.: LLM meets bounded model checking: neuro-symbolic loop invariant inference. In: Proceedings of the 39th IEEE/ACM International Conference on Automated Software Engineering, pp. 406–417. ACM (2024). https://doi.org/10.1145/3691620.3695014

Engineering Trustworthy Automation: Design Principles and Evaluation for AutoML Tools for Novices

Jarne Thys[(✉)] [iD], Davy Vanacken [iD], and Gustavo Rovelo Ruiz [iD]

UHasselt, Hasselt University, Digital Future Lab—Flanders Make, Diepenbeek, Belgium
{jarne.thys,davy.vanacken,gustavo.roveloruiz}@uhasselt.be

Abstract. AutoML systems targeting novices often prioritize algorithmic automation over usability, leaving gaps in users' understanding, trust, and end-to-end workflow support. To address these issues, we propose an abstract pipeline that covers data intake, guided configuration, training, evaluation, and inference. To examine the abstract pipeline, we report a user study where we assess trust, understandability, and UX of a prototype implementation. In a 24-participant study, all participants successfully built their own models, UEQ ratings were positive, yet experienced users reported higher trust and understanding than novices. Based on this study, we propose four design principles to improve the design of AutoML systems targeting novices: (P1) support first-model success to enhance user self-efficacy, (P2) provide explanations to help users form correct mental models and develop appropriate levels of reliance, (P3) provide abstractions and context-aware assistance to keep users in their zone of proximal development, and (P4) ensure predictability and safeguards to strengthen users' sense of control.

Keywords: AutoML · Large Language Models · Transformers · Text Classification · Conversational Assistant

1 Introduction

Novices are increasingly interested in training AI models, both professionally and personally. Professionally, AI's growing impact across industries creates a need for workers to develop AI skills to remain competitive in evolving job markets [4]. Personally, participants in studies expressed pride in contributing to the improvement of AI, highlighting the emotional connection and personal interest in AI development [38]. Additionally, the potential for complementary performance between humans and AI systems motivates engagement. When humans and AI collaborate effectively, their combined performance can exceed what either could achieve alone [17]. This potential becomes especially relevant in organizational settings where AI systems are deployed alongside domain experts, not only to support task execution but also to help transfer expert knowledge to less experienced users [30]. As a result, such collaboration provides a compelling reason for novices to invest in developing AI competencies, both to enhance their own performance and to contribute to human-AI teams.

Meanwhile, new advanced AI/ML architectures can be applied to the specialized domains of those users. Professors could automatically grade open-ended questions using Transformer-based classifiers, chemists could simulate reactions using molecular graphs with Graph Neural Networks, and clinicians could model continuous physical processes using Liquid Neural Networks. However, despite technological progress, the practical adoption of these advanced architectures by AI/ML novices remains limited. Deploying advanced architectures typically requires fluency with programming, familiarity with AI/ML frameworks, and careful attention to error-prone details such as dataset validation, model configuration, and training orchestration. As a result, domain experts without AI/ML expertise often struggle to leverage these models for their own data [25].

Efforts to lower these barriers include AutoML platforms and no-code/low-code AI systems, which automate individual steps of the pipeline and expose model training through graphical interfaces. While such tools improve accessibility, they leave critical gaps for novices. They often emphasize algorithm optimization and offer little support for end-to-end workflows (e.g., dataset validation, feature selection, inference setup) [36]. They also rarely evaluate how their abstractions affect user trust and interpretability, focusing on technical aspects and leaving open questions about how novices experience such systems [21]. To address these challenges, our contributions are threefold:

1. **An abstract AutoML Pipeline for Novices.** We propose an abstract end-to-end pipeline designed to support novices that links data intake, configuration, training, evaluation, and inference.
2. **Evaluation of the Pipeline via a Prototype for Novice-Oriented Workloads.** A 24-participant study tests (i) end-to-end feasibility across various datasets and tasks using Transformer-based text classification, and (ii) robustness of the training and inference via metadata-driven pipelines.
3. **Design Principles for AutoML Tools for Novices.** Based on the study results and relevant theories, we present four design principles for future AutoML systems.

2   Related Work

A machine learning pipeline is the end-to-end process that transforms raw data into a deployable model. The pipeline includes data collection and cleaning, feature engineering, model selection, training, evaluation, and deployment. Each stage introduces challenges, such as missing or biased data, feature drift as real-world conditions evolve, and concerns about interpretability and reproducibility, which complicate the process of building a reliable model. Beyond these technical hurdles, many domain experts still struggle to train AI models, as doing so typically requires programming fluency and familiarity with ML frameworks. These barriers have motivated the development of AutoML tools that automate or abstract parts of the pipeline. We review three topics relevant to novice-oriented AutoML systems: (i) AutoML approaches that automate model selection and optimization but often overlook pipeline orchestration and user understanding; (ii) no- and low-code platforms that broaden access while sometimes obscuring key decisions or limiting diagnostic support; and (iii) AI assistants and contextual help systems that promise step-aware guidance but vary in reliability and depth of integration.

AutoML Solutions and Challenges. AutoML is an alternative for users with limited technical knowledge. These platforms automate various components of the machine learning pipeline to lower the entry barrier for novices. Notable systems include Auto-sklearn, TPOT, and commercial platforms like Google AutoML, DataRobot, and Azure Machine Learning Studio [24,28]. These systems employ different search strategies, such as Bayesian optimization combined with meta-learning [29], genetic programming [24], and reinforcement learning [9], to automatically generate and optimize ML pipelines. However, AutoML platforms face several limitations that prevent them from achieving their promise of fully automated machine learning. First, much of AutoML research has focused on isolated parts of the ML pipeline, such as preprocessing or hyperparameter optimization, rather than full end-to-end workflows, which often makes these methods difficult to apply without expert oversight [28]. Second, the search process for high-performance models can be extremely slow, taking minutes to hours, which affects system interactivity and necessitates asynchronous communication channels [12]. Consequently, studies show that AutoML users are still primarily expert data scientists, and the tools require skilled users [7,13].

No-Code/Low-Code AI Platforms. To cater to a non-expert audience, different no-code and low-code platforms have been developed. These platforms represent the most accessible tier of interactive machine learning systems, designed to enable users with little to no programming expertise to create and deploy AI models through graphical interfaces [23]. They typically provide user-friendly interfaces with tools that can automatically handle data processing tasks, such as finding missing data, identifying incorrect labels, and selecting desired data subsets [23]. Recent advances leverage Large Language Models (LLMs) to create conversational interfaces that can iteratively extract user requirements and provide real-time guidance throughout the model-building process [22,31]. These natural language interfaces show particular promise for bridging technical knowledge gaps, allowing users across different expertise levels to successfully complete complex machine learning tasks [37].

AI Assistants in Complex Software Systems. The use of AI assistants in complex systems addresses a key limitation: while current techniques handle simple tasks effectively, they struggle to generalize to conversational interfaces that help humans solve complex problems through interaction with AI reasoning systems [2]. In practice, this spans domains from software engineering, where assistants ask clarifying questions and generate code [27], to business operations, where intelligent task assistants execute processes via multi-agent orchestration that maps natural-language requests to executable sequences of operations connected to back-end services [5,16]. These Conversational Agentic Systems combine the conversational capabilities of LLMs with structured function calls and typically require specialized dialogue fine-tuning to preserve coherence over extended workflows [26]. Such assistants can be further extended by taking context into account when providing support.

Contextual Help and Guidance Systems. The effectiveness of contextual help systems depends heavily on their ability to predict and respond to user satisfaction and engagement in real-time, particularly in open-domain conversations without clearly defined goals [6]. Conversational interfaces offer notable advantages over traditional WIMP (Windows, Icons, Menus, and Pointers) interfaces by providing natural and familiar interaction methods, flexible accommodation of diverse user requests, and anthropomorphic features that help attract attention and gain trust, yet they continue to face significant challenges in processing natural language expressions and managing complex conversation situations [15,35]. LLMs can be used to address the limitations of previous systems in complex conversations. They have significantly enhanced contextual guidance capabilities, with multi-turn conversational prompting making LLMs more responsive and proficient in handling complex queries and extended discussions. These systems now demonstrate improved fluidity and relevance in interactions, making them more engaging and helpful across applications ranging from customer service to therapy bots [8].

3   Abstract Pipeline for AutoML Tools for Novices

Tools that create abstractions from advanced AI/ML architectures (e.g., Transformers, Graph Neural Networks, Liquid Neural Networks) can empower non-experts, but only if the abstraction is designed for usability and understanding; automation alone is not enough [21]. In this section, we introduce an abstract pipeline of an AutoML tool to support novices. The full pipeline is illustrated in Fig. 1.

Data Intake and Upfront Safeguards. The pipeline begins with data ingestion and validation mechanisms. Automated data processing and feature engineering can streamline data preparation tasks [23]. Domain experts can contribute knowledge relevant to data preprocessing and feature engineering [21], safeguards at this stage address data quality and potential biases [7].

Key Model Parameter Configuration. The user should only be exposed to parameters that fundamentally change the model and where their domain knowledge is relevant [21], such as the selection of input features and the objective of the model. Other parameters (e.g., model, optimizer, learning

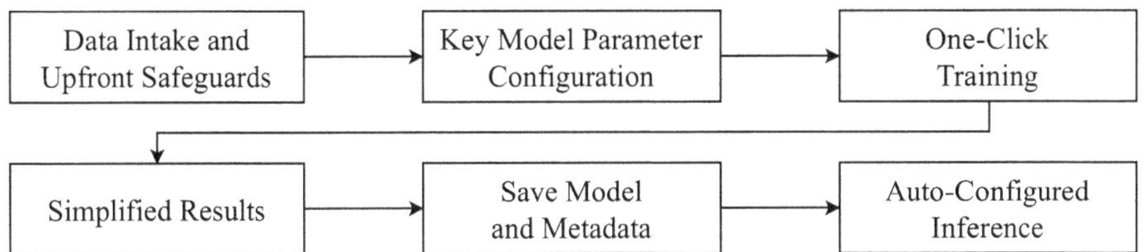

Fig. 1. An abstract, end-to-end pipeline for AutoML tools targeting novice users. It begins with Data Intake and Upfront Safeguards, then narrows decisions to Key Model Parameter Configuration before One-Click Training. After training, Simplified Results explain performance in simplified terms, while Save Model and Metadata preserves an inference "contract." Finally, Auto-Configured Inference uses that contract to generate input fields and outputs consistently.

rate) can either be auto-configured by the tool or be a static, robust baseline that can deliver sufficient performance for most cases. We argue that peak performance is not a requirement for experimentation.

One-Click Training. After the key parameters have been configured, the tool should handle all configuration without any user input. All parts of the ML pipeline (e.g., train-val-test splitting, handling missing values, converting string labels to integers) should be handled automatically.

Simplified Results. Transparency and interpretability affect user trust in AutoML tools, with interpretability identified as a key user requirement [21]. Result presentation should accommodate different stakeholder needs, from domain experts to ML practitioners [13,37].

Save Model and Metadata. This stage addresses model checkpointing, saving, and inference through metadata. All relevant training data that can be carried over to inference should be saved, as this metadata can be used as type hints for the user to correctly use the trained model and for the tool to configure a trained model for inference without any user input.

Auto-Configured Inference. The final stage provides automated deployment with provisions for human oversight in high-stakes applications. Inference should automate technical aspects of model serving [21] based on the metadata saved in the previous stage. The design should combine human expertise and AutoML capabilities, particularly for strategic decisions, ethical considerations, and domain-specific requirements [7,21].

4 NovaClass: Applying the Abstract Pipeline to Transformer-Based Classification

To put the abstract pipeline into practice, the following section details Nova-Class, our novice-friendly automation prototype, which aims to lower the entry barrier for novices who wish to fine-tune Transformer models for text classification tasks. We use supervised text classification as it has many use cases (e.g., grading, spam detection, emotion classification), and performance has taken a

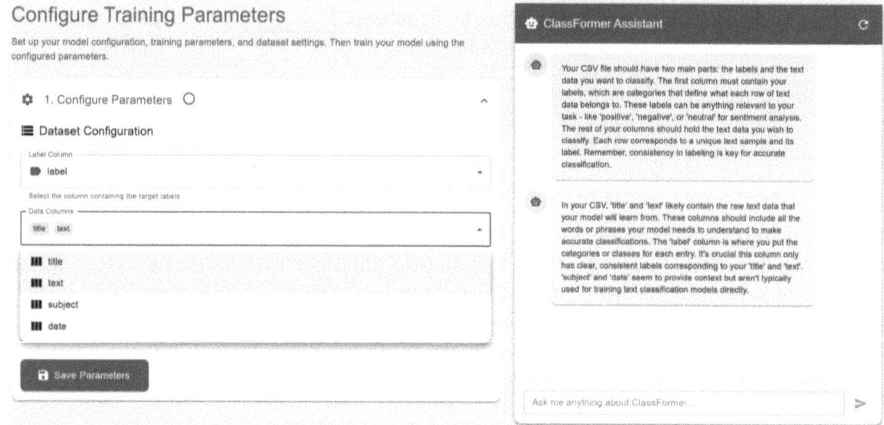

Fig. 2. The NovaClass interface for configuring model parameters. On the left, users select which columns provide the input text and which column contains the target labels, choices that determine how the classifier is trained. On the right, an integrated contextual assistant gives real-time explanations and guidance to support novices during configuration.

leap forward with the introduction of the Transformer architecture. The use of Transformers, however, still requires in-depth knowledge to set up a suitable pipeline.

Data Intake and Analysis. Users upload a single CSV for which NovaClass highlights column types, missing rows, and label balance. Furthermore, users can inspect class distributions and statistical analysis for numerical columns. Finally, the users can inspect the first ten rows of the dataset.

Automatic Classifier Generator. As illustrated in Fig. 2, we only expose the decisions that novices are expected to understand (input columns, target labels) and run a reproducible pipeline with safe defaults, aiming to produce a working baseline on the first attempt. Model metadata are saved to eliminate train–inference mismatches and promote consistent behavior.

Cascade Classification Strategy. To enable novices to use advanced classification strategies, NovaClass integrates a one-toggle cascaded classification option. As illustrated in Fig. 3, NovaClass decomposes multi-class prediction tasks into a sequence of simpler binary decisions, supporting the one-click training even with advanced classification strategies. Instead of training a single model to discriminate across all categories simultaneously, the cascade arranges multiple binary classifiers in a hierarchical structure. This sequential breakdown can reduce label imbalance and limit the risk of confusion between adjacent categories, aiming to improve recall for underrepresented classes and precision for high-confidence ones [32].

Inference of the Trained Model. The inference view uses metadata (e.g., strategy, encoders, label order), to prevent mismatches between training and

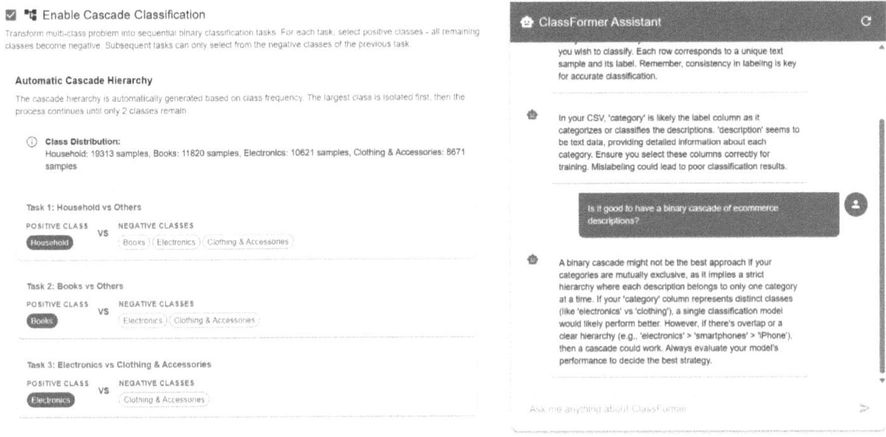

Fig. 3. The NovaClass interface for cascade classification (left), supported by the integrated contextual assistant (right). Instead of training a single multi-class model, the system automatically builds a hierarchy of binary classifiers. This cascade is generated based on class frequency, automatically addressing label imbalance and making it easier for novices to apply advanced classification strategies without additional configuration.

inference. Outputs present both the predicted label confidence and the full class-probability distribution to support quick plausibility checks.

Conversational Assistant. A context-aware assistant provides simplified explanations of metrics (e.g., accuracy, F1, recall) and suggests next steps appropriate to the user's stage, aiming to reduce jargon and decision burden. We use IBM's Granite 3.3 8B[1] as, in our testing, it adhered best to instructions, has a context window of 128K to be able to handle larger contexts, and is small enough to run locally, as we aim to provide a secure environment for users with confidential and/or sensitive data.

5   User Testing

5.1   Methodology

We conducted a study designed to evaluate how users interact with NovaClass across three different tasks of varying complexity. **Task 1** is a binary classification task that assesses whether users can successfully configure a simple text classifier with multiple candidate input fields. This provides a baseline of how effectively the guided dataset configuration supports novices in making informed feature and label selections. **Task 2** is a cascade classification task that examines whether participants can set up a hierarchical pipeline and allows us to test

---

[1] https://www.ibm.com/new/announcements/ibm-granite-3-3-speech-recognition-refined-reasoning-rag-loras.

the system's ability to introduce more advanced classification strategies in an approachable manner. **Task 3** is a diagnosis task that focuses on result interpretation, testing whether the available tools help users identify performance issues due to class imbalance in the dataset. We ordered the tasks using a balanced Latin square to mitigate carry-over effects. For the study, we used datasets to classify fake news[2], e-commerce product descriptions[3], and student grades[4].

The demographic information for the study was collected through a questionnaire at the beginning of the study. Participants are asked to provide their age group, highest degree obtained, and self-described gender identity. In addition, the questionnaire captured participants' prior exposure to artificial intelligence and machine learning by asking whether they had used AutoML tools before and whether they had ever trained or fine-tuned models. Finally, we asked the participants to complete the PAILQ-6 (Perceived Artificial Intelligence Literacy) questionnaire [14].

To evaluate NovaClass, and in extension our design principles, we base our questionnaire on the study by Drozdal et al. [10]. It targets trust and understandability through 14 items, of which two[5] were removed as they are only applicable to systems where multiple model architectures can be trained and compared. Responses are collected on a 5-point Likert scale, supplemented by a binary question asking the participant whether they would deploy their trained model in the real world. Additionally, we asked the participants to complete the User Experience Questionnaire (UEQ) [19].

5.2   Results

Our study included 24 participants (18 male, 6 female), with a majority of them in age groups 18–24 ($n = 8$) and 25–34 ($n = 9$). Most of them ($n = 22$) indicated that they had never used AutoML tools before. To investigate differences between novice and experienced users, we created two groups: users who had previously trained ML models themselves (experienced users, $n = 16$) and users who had never trained a model before (novices, $n = 8$). In the experienced group, most users ($n = 13$) had trained deep neural models before, while 6 users had fine-tuned models before. Four users indicated that they had trained models as part of a course, but not outside of the course. Using a Mann-Whitney U test, we found a significant effect on the Perceived Artificial Intelligence Literacy scores between the two groups. The mean ranks of the experienced group and novice group were 15.06 and 7.38, respectively ($U = 23$, $Z = 2.52$, $p < 0.05$, $r = 0.51$). The medians of participants' mean 7-point Likert ratings were 5.5 and 4.4 for the experienced group and novice group, respectively.

---

[2] https://www.kaggle.com/datasets/clmentbisaillon/fake-and-real-news-dataset.
[3] https://doi.org/10.5281/zenodo.3355822.
[4] https://www.kaggle.com/datasets/mahmoudelhemaly/students-grading-dataset.
[5] "I understand how estimators are selected" and "I understand the differences between the generated models".

Task Performance

Task 1. All participants were able to train a working classifier model to predict fake news articles. We asked the participants to rate their confidence level both for training and using a binary classification model on a 7-point Likert scale (Extremely unconfident - Extremely confident). 21 participants (87.5%) indicated at least some level of confidence in training the model, while 3 participants (12.5%) indicated they were extremely unconfident.

Task 2. All participants were able to correctly train a cascade classification model to classify e-commerce descriptions. Participants were asked to determine which task in the cascade performed the worst and which tools they used to reach that conclusion. 22 participants (91.7%) were able to correctly identify the lowest performing task. We asked participants to indicate which NovaClass tools they used to identify the lowest-performing task. The most-used tool was the classification report ($n = 14$), closely followed by the conversational assistant ($n = 13$).

Task 3. 17 participants (70.8%) were able to correctly identify the label imbalance in the dataset. The participants were asked to rate whether they thought they correctly identified the issue on a 5-point Likert scale (Definitely not - Definitely yes). 18 participants (75%) reported at least some level of certainty of correctly identifying the problem. We asked participants to indicate which NovaClass tools they used to diagnose the issue. The most-used feature was the conversational assistant ($n = 15$), closely followed by the data analysis ($n = 14$) and the confusion matrix ($n = 13$).

User Experience, Trust, and Understandability

User Experience. Participants evaluated the system positively across all six UEQ dimensions (scale -3 to +3, full results in Table 1, visualized in Fig. 4). The highest ratings were observed for efficiency, attractiveness, and perspicuity, indicating that the tool was perceived as effective, appealing, and relatively easy to understand. Dependability and stimulation also received high ratings, suggesting that participants considered the system reliable and engaging. Novelty was evaluated somewhat lower in comparison to the other dimensions. We note, however, that the internal consistency for some scales was limited. Specifically, Cronbach's Alpha was below the commonly accepted threshold of 0.7 for Efficiency ($\alpha = 0.61$) and Dependability ($\alpha = 0.49$). This may indicate heterogeneous responses or that participants understood the items differently.

Trust and Understandability. Using a Mann-Whitney U test, we found a significant effect of experience on the average trust and understandability scores between the two groups. The medians of participants' mean 5-point Likert ratings were 4.05 and 3.64 for the experienced group and novice group, respectively.

Table 1. UEQ results for the six standard scales (Attractiveness, Perspicuity, Efficiency, Dependability, Stimulation, Novelty) on the -3 to +3 evaluation range (higher is better). For each scale, we report the sample Mean, Std. Dev. across participants ($n = 24$), the half-width of the 95% confidence interval, and the corresponding lower/upper bounds around the mean. We note that Cronbach's Alpha was below the commonly accepted threshold of 0.7 for Efficiency ($\alpha = 0.61$) and Dependability ($\alpha = 0.49$).

Scale	Mean	Std. Dev.	Confidence	Confidence interval	
				Lower	Upper
Attractiveness	1.778	0.677	0.271	1.507	2.049
Perspicuity	1.635	1.096	0.438	1.197	2.074
Efficiency	1.917	0.658	0.263	1.653	2.180
Dependability	1.375	0.634	0.254	1.121	1.629
Stimulation	1.458	0.743	0.297	1.161	1.756
Novelty	1.052	0.831	0.332	0.720	1.384

The mean ranks of the experienced group and novice group were 14.91 and 7.69, respectively ($U = 25.5$, $Z = 2.37$, $p < 0.05$, $r = 0.48$). Next, we investigated each question from the questionnaire individually using a Mann-Whitney U test. We found a significant effect of experience between the two groups for the following questions:

- **"I understand the tool."** The median 5-point Likert ratings were 4.5 and 4 for the experienced group and novice group, respectively. The mean ranks of the experienced group and novice group were 14.75 and 8, respectively ($U = 28$, $Z = 2.43$, $p < 0.05$, $r = 0.50$).
- **"I understood the tool's overall process."** The median 5-point Likert ratings were 5 and 4 for the experienced group and novice group, respectively. The mean ranks of the experienced group and novice group were 14.43 and 8.63, respectively ($U = 33$, $Z = 2.10$, $p < 0.05$, $r = 0.43$).
- **"I understand the data."** The median 5-point Likert ratings were 5 and 4 for the experienced group and novice group, respectively. The mean ranks of the experienced group and novice group were 15.22 and 7.06, respectively ($U = 20.5$, $Z = 2.94$, $p < 0.05$, $r = 0.60$).
- **"I understand the model evaluation metrics."**: The median 5-point Likert ratings were 5 and 2.5 for the experienced group and novice group, respectively. The mean ranks of the experienced group and novice group were 15.03 and 7.44, respectively ($U = 23.5$, $Z = 2.74$, $p < 0.05$, $r = 0.56$).

When asked whether they would deploy models trained with NovaClass, 17 participants (70.8%) answered yes. The most recurring reason ($n = 10$) was a variant of "high accuracy" or "high F1-score," but experienced users would also give more detailed feedback. For example, one experienced user mentioned "There were no fake articles mentioned as real, which indicates a low chance of

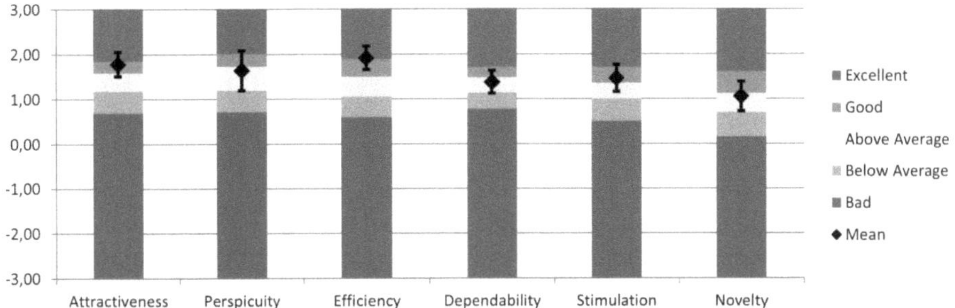

Fig. 4. Distribution of UEQ responses and comparison to the UEQ benchmark. The colored background bars are reference bands from the UEQ benchmark: Bad (bottom 25%), Below average (25–50th percentile), Above average (50–75th percentile), Good (75–90th percentile), and Excellent (top 10%). For each scale (range -3 to +3), the black diamond and whiskers show our sample mean and 95% CI; their position against the bands indicates the benchmark class of our product. Overall, ratings are positive, highest for Efficiency, Attractiveness, and Perspicuity; positive but more moderate for Dependability and Stimulation; and comparatively lower for Novelty.

the worst case scenario," referencing the precision and recall in the classification report. Out of the 7 participants who chose not to deploy the models, there were mainly issues about the transparency of the tool and/or the models. Participants, for example, mentioned "I am unsure because I do not know what the model bases itself on to make a decision," "No overview of how the data is processed, so not reliable to implement in production," and "I can not know for sure if the test dataset that is used for the evaluation is biased towards the training data or not." These comments are in line with the differences we found for questions about the understanding of the data and evaluation metrics.

The low understanding of the tool and overall process can also be linked to observations made, especially during Task 2. Many users, novice and experienced, would enable the cascade classification as per the instructions, yet paid no attention to the system's explanation about what would happen, how the data would be processed, or what the final binary classifiers would look like.

5.3   Discussion

This study set out to understand whether a guided, end-to-end pipeline can make Transformer fine-tuning accessible to novices, and how design choices shape trust and understanding. Overall, participants completed the core tasks successfully: everyone trained a working binary classifier (Task 1) and cascaded classifier (Task 2), and most participants correctly diagnosed class imbalance in the analysis task (Task 3). These outcomes suggest that NovaClass's training pipeline and metadata-driven inference help reduce the kinds of configuration errors that commonly block novice progress. The positive UEQ ratings for efficiency, attractiveness, and perspicuity reinforce this, indicating that participants perceived the tool as effective and comprehensible in practice.

At the same time, the results reveal a persistent gap between novice and experienced users. Experienced participants reported higher trust and understandability scores, with significant group differences on items about understanding the tool, the overall process, the data, and evaluation metrics. These differences likely reflect not only prior exposure to ML concepts but also how users interpret model feedback: those with prior experience may map metrics and visualizations to mental models more readily, whereas novices need more scaffolding to connect outputs to actionable insight. The fact that 17 of 24 participants would deploy their models, with high accuracy or F1-score as the most common reason, shows that many participants equate performance indicators with deployability, while the seven who hesitated emphasized the black-box nature of NovaClass and the trained models. Both reactions underscore the need to couple performance reporting with transparent, digestible explanations of model behavior and limits. Even so, novices did not evaluate the system negatively: their mean trust/understandability scores remained mostly positive, aligning with the broadly positive UEQ results and high task success rates. This suggests the gap reflects relative differences rather than dissatisfaction.

The conversational assistant played a central role in sense-making across tasks. Participants relied on the conversational assistant to select input features (Task 1), identify the weak cascade stage (Task 2), and diagnose imbalance (Task 3), often as much as or more than static tools like the classification report or confusion matrix. This pattern suggests that context-aware, real-time guidance can assist users by translating results into plain-language cues tied to the user's current step. However, dependence on the assistant also increases the cost of occasional hallucinations or imprecise explanations, which several participants encountered.

Defaults and automation enabled novices to quickly reach "first model success," while the cascade option let all participants experience a more advanced strategy without extra configuration. Yet the uniform interface likely underserved both ends of the spectrum: novices who would benefit from additional safeguards and predictive hints, and experienced users who asked for deeper controls and richer diagnostics.

6   Design Principles for AutoML Tools for Novices

We translate our findings into actionable guidance for developers of AutoML tools aimed at novices. We combine the abstract pipeline, relevant theories, and insights from our study into four principles that aim to raise self-efficacy, calibrate trust, and preserve user control while still accommodating expert needs.

P1   First-Model Success to Raise Self-Efficacy
Ensure a working baseline on the first attempt so novices experience immediate success. Early mastery experiences measurably increase self-efficacy and persistence; a near-guaranteed first win is a reliable way to raise confidence for later tasks [3]. To help first-model success, we propose the following implementation guidelines:

– Ship safe defaults with one-click training and make failure hard: validate inputs up front (e.g., schema, labels, missing values), choose conservative parameters, and automatically handle actions like preprocessing, label order, checkpointing, and recovery from common errors.
– Provide real-time feedback (e.g., "Model trained", "F1-score=...") combined with next-step nudges (e.g., "Try other input features").
– Combine ease of use with advanced functionality: enable advanced features (like cascade classification) through single toggles that maintain the one-click training workflow.
– Preserve first-model success across expertise levels: Regardless of interface adaptation or user expertise, maintain the goal of first-model success through consistent defaults and failsafe mechanisms. Dataset-agnostic strategies and advanced techniques should be deployable with one click, ensuring novices benefit from stronger baselines while experts retain inspection and override capabilities.

P2   Explanations to Create Mental Models and Appropriate Reliance
Pair metrics with simplified explanations so users understand what the model did and how well it performed. Explanatory debugging improves users' mental models [18], and appropriate reliance requires transparency beyond raw scores [20]; both can reduce over- and under-trust. Our study revealed that while 71% of participants would deploy their models based on high accuracy scores, those who hesitated cited concerns about not knowing "what the model bases itself on to make a decision" and lack of "overview of how the data is processed." To help create mental models and appropriate reliance, we propose the following implementation guidelines:

– Add tiered explanations for key metrics. When novice users inspect a score (e.g., F1-score, precision, recall) or a visualization (e.g., confusion matrix, ROC curve), show a short tooltip describing what the metric measures, how to interpret high or low values, and one simple suggestion for improvement. When experienced users inspect performance, the system should provide more in-depth and advanced metrics.
– Add appropriate reliance cues next to each metric (e.g., "High F1-score with low minority-class recall risks under-serving class Y").
– Make system processes (e.g., data preprocessing, train-test splitting, validation steps) transparent and inspectable. Users need to be able to understand the system's entire process.
– Implement tool-augmented conversational approaches [33] that enable users to interactively explore model decisions through natural dialogue. These systems combine large language models with explanatory tools, allowing users to ask questions like "Why was this classified as X?" or "Show me examples where the model confuses class A and B," and receive context-aware, data-driven responses. Such conversational explainability improves transparency and user understanding by supporting follow-up questions and deeper exploration of model behavior.

P3   Abstractions and Context-Aware Assistance to Support the Zone of Proximal Development

Offer targeted guidance through abstracted interfaces and context-aware assistance so novices operate within their zone of proximal development [34]. Abstractions and context-aware support turn opaque steps (e.g., configuration, evaluation, inference) into guided actions, helping users perform just beyond their independent ability. Our study found significant differences in understanding between novice and experienced users, with novices scoring lower on understanding the tool, overall process, data, and evaluation metrics. Additionally, many users enabled cascade classification features without engaging with explanations. To help create better abstractions and context-aware assistance, we propose the following implementation guidelines:

- Implement interfaces that adapt to ML experience using brief proficiency checks to keep users in their zone of proximal development.
- Embed a context-aware assistant that can help check for common problems (e.g., imbalance, too little data) and suggest the next action. To interpret data, address reliability concerns through hybrid approaches that combine LLM flexibility with templated responses to reduce hallucinations (specifically of performance numbers) while maintaining natural interaction.
- Implement advanced, dataset-agnostic ML strategies (e.g., model multiplicity [11]) that require no configuration (one-click training) but allow expert inspection and/or configuration.

P4   Predictability and Safeguards to Strengthen Perceived Control

Make the system predictable and safe with metadata-driven UIs and strict safeguards that prevent train-inference mismatches. In the theory of planned behavior, perceived behavioral control is a key driver of intention. Clear constraints, validation, and consistency increase users' felt control [1]. To help create more predictability and better safeguards, we propose the following implementation guidelines:

- Generate inference UIs directly from training metadata (e.g., schema, label order, input features).
- Provide pre-flight checks (e.g., schema, missing values, label coverage) and actionable errors with safe fallbacks.
- Implement validation and constraints appropriate to the user's expertise level, with novices receiving more protective guardrails and experienced users having more flexibility.

7   Conclusion

This paper contributes an **abstract end-to-end pipeline covering data intake, configuration, training, evaluation, and inference**. The emphasis of this pipeline is on reliability for novices and user understanding rather than algorithmic optimization for peak performance. We used a **prototype implementation** to examine the abstract pipeline in practice. In a **24-participant study**, all participants successfully trained a binary and cascaded classifier. **User-experience ratings were positive**, and 17 participants reported they would deploy their models. However, **experienced users reported higher trust and understanding** than novices, and several participants raised transparency concerns.

Based on these findings and relevant theories, we propose a set of **four design principles for AutoML tools for novices**: (P1) first-model success, (P2) explanations that support correct mental models and appropriate reliance, (P3) abstractions with context-aware assistance, and (P4) predictability through safeguards and metadata. Based on these design principles, future AutoML tools could (i) combine LLM-driven guidance with templated summaries to reduce hallucinations and maintain step awareness, (ii) implement expertise-adaptive scaffolding that preserves first-run defaults while progressively revealing controls, and (iii) develop accessible explainability that links metrics and visualizations to concise, actionable narratives.

Acknowledgments. This work was supported by the Special Research Fund (BOF) of Hasselt University (BOF24OWB28). This research was made possible with support from the MAXVR-INFRA project, a scalable and flexible infrastructure that facilitates the transition to digital-physical work environments. The MAXVR-INFRA project is funded by the European Union - NextGenerationEU and the Flemish Government.

Disclosure of Interests. The authors have no competing interests to declare that are relevant to the content of this work.

References

1. Ajzen, I.: From intentions to actions: a theory of planned behavior. In: Kuhl, J., Beckmann, J. (eds.) Action Control. SSSP Springer Series in Social Psychology, pp. 11–39. Springer, Heidelberg (1985). https://doi.org/10.1007/978-3-642-69746-3_2
2. Allen, J., Galescu, L., Teng, C.M., Perera, I.: Conversational agents for complex collaborative tasks. AI Mag. **41**(4), 54–78 (2020). https://doi.org/10.1609/aimag.v41i4.7384
3. Bandura, A.: Self-efficacy: toward a unifying theory of behavioral change. Psychol. Rev. **84**(2), 191–215 (1977). https://doi.org/10.1037/0033-295X.84.2.191
4. Brawner, K., Wang, N., Nye, B.: Teaching artificial intelligence (AI) with AI for AI applications. Int. FLAIRS Conf. Proc. **36** (2023). https://doi.org/10.32473/flairs.36.133388

5. Chakraborti, T., Agarwal, S., Khazaeni, Y., Rizk, Y., Isahagian, V.: D3BA: a tool for optimizing business processes using non-deterministic planning. In: Del Río Ortega, A., Leopold, H., Santoro, F.M. (eds.) Business Process Management Workshops, pp. 181–193. Springer International Publishing, Cham (2020). https://doi.org/10.1007/978-3-030-66498-5_14

6. Choi, J.I., Ahmadvand, A., Agichtein, E.: Offline and online satisfaction prediction in open-domain conversational systems. In: Proceedings of the 28th ACM International Conference on Information and Knowledge Management, pp. 1281–1290. CIKM '19, Association for Computing Machinery (2019).https://doi.org/10.1145/3357384.3358047

7. Crisan, A., Fiore-Gartland, B.: Fits and starts: enterprise use of AutoML and the role of humans in the loop. In: Proceedings of the 2021 CHI Conference on Human Factors in Computing Systems, pp. 1–15. CHI '21, Association for Computing Machinery (2021). https://doi.org/10.1145/3411764.3445775

8. Ding, B., et al.: Data augmentation using LLMs: data perspectives, learning paradigms and challenges. In: Ku, L.W., Martins, A., Srikumar, V. (eds.) Findings of the Association for Computational Linguistics: ACL 2024, pp. 1679–1705. Association for Computational Linguistics (2024). https://doi.org/10.18653/v1/2024.findings-acl.97

9. Drori, I., et al.: AlphaD3M: machine learning pipeline synthesis. In: ICML AutoML Workshop (2021)

10. Drozdal, J., et al.: Trust in AutoML: exploring information needs for establishing trust in automated machine learning systems. In: Proceedings of the 25th International Conference on Intelligent User Interfaces, pp. 297–307. IUI '20, Association for Computing Machinery (2020). https://doi.org/10.1145/3377325.3377501

11. Eerlings, G., Vanbrabant, S., Liesenborgs, J., Rovelo Ruiz, G., Vanacken, D., Luyten, K.: AI-spectra: a visual dashboard for model multiplicity to enhance informed and transparent decision-making. In: Zaina, L., et al. (ed.) Engineering Interactive Computer Systems. EICS 2024 International Workshops, pp. 55–73. Springer Nature Switzerland, Cham (2025). https://doi.org/10.1007/978-3-031-91760-8_5

12. Egelé, R., Guyon, I., Vishwanath, V., Balaprakash, P.: Asynchronous decentralized bayesian optimization for large scale hyperparameter optimization. In: 2023 IEEE 19th International Conference on e-Science (e-Science), pp. 1–10 (2023). https://doi.org/10.1109/e-Science58273.2023.10254839

13. Feng, K.J.K., Mcdonald, D.W.: Addressing UX practitioners' challenges in designing ML applications: an interactive machine learning approach. In: Proceedings of the 28th International Conference on Intelligent User Interfaces, pp. 337–352. IUI '23, Association for Computing Machinery (2023). https://doi.org/10.1145/3581641.3584064

14. Grassini, S.: A psychometric validation of the PAILQ-6: perceived artificial intelligence literacy questionnaire. In: Proceedings of the 13th Nordic Conference on Human-Computer Interaction, pp. 1–10. NordiCHI '24, Association for Computing Machinery (2024). https://doi.org/10.1145/3679318.3685359

15. Grudin, J., Jacques, R.: Chatbots, humbots, and the quest for artificial general intelligence. In: Proceedings of the 2019 CHI Conference on Human Factors in Computing Systems, pp. 1–11. CHI '19, Association for Computing Machinery (2019). https://doi.org/10.1145/3290605.3300439

16. He, J., Piorkowski, D., Muller, M., Brimijoin, K., Houde, S., Weisz, J.: Rebalancing worker initiative and AI initiative in future work: four task dimensions. In: Proceedings of the 2nd Annual Meeting of the Symposium on Human-Computer Interaction for Work, pp. 1–16. CHIWORK '23, Association for Computing Machinery (2023). https://doi.org/10.1145/3596671.3598572
17. Hemmer, P., Schemmer, M., Kühl, N., Vössing, M., Satzger, G.: On the effect of information asymmetry in human-AI teams. In: CHI Conference on Human Factors in Computing Systems (CHI '22), Workshop on Human-Centered Explainable AI (HCXAI) (2022). arXiv. https://doi.org/10.48550/ARXIV.2205.01467
18. Kulesza, T., Burnett, M., Wong, W.K., Stumpf, S.: Principles of explanatory debugging to personalize interactive machine learning. In: Proceedings of the 20th International Conference on Intelligent User Interfaces, pp. 126–137. ACM (2015). https://doi.org/10.1145/2678025.2701399
19. Laugwitz, B., Held, T., Schrepp, M.: Construction and evaluation of a user experience questionnaire. In: Holzinger, A. (ed.) HCI and Usability for Education and Work, pp. 63–76. Springer, Heidelberg (2008). https://doi.org/10.1007/978-3-540-89350-9_6
20. Lee, J.D., See, K.A.: Trust in automation: designing for appropriate reliance. Hum Fact.: J. Hum. Fact. Ergon. Soc. **46**(1), 50–80 (2004). https://doi.org/10.1518/hfes.46.1.50_30392
21. Lindauer, M., et al.: Position: a call to action for a human-centered AutoML paradigm. In: Proceedings of the 41st International Conference on Machine Learning. ICML'24, vol. 235, pp. 30566–30584. JMLR.org (2024)
22. Luo, D., Feng, C., Nong, Y., Shen, Y.: AutoM3L: an automated multimodal machine learning framework with large language models. In: Proceedings of the 32nd ACM International Conference on Multimedia, pp. 8586–8594. MM '24, Association for Computing Machinery (2024). https://doi.org/10.1145/3664647.3680665
23. Mumuni, A., Mumuni, F.: Automated data processing and feature engineering for deep learning and big data applications: a survey. J. Inf. Intell. **3**(2), 113–153 (2025). https://doi.org/10.1016/j.jiixd.2024.01.002
24. Olson, R.S., Bartley, N., Urbanowicz, R.J., Moore, J.H.: Evaluation of a tree-based pipeline optimization tool for automating data science. In: Proceedings of the Genetic and Evolutionary Computation Conference 2016, pp. 485–492. GECCO '16, Association for Computing Machinery (2016). https://doi.org/10.1145/2908812.2908918
25. Paleyes, A., Urma, R.G., Lawrence, N.D.: Challenges in deploying machine learning: a survey of case studies. ACM Comput. Surv. **55**(6), 114:1-114:29 (2022). https://doi.org/10.1145/3533378
26. Robino, G.: Conversation routines: a prompt engineering framework for task-oriented dialog systems (2025). https://doi.org/10.48550/arXiv.2501.11613
27. Ross, S.I., Martinez, F., Houde, S., Muller, M., Weisz, J.D.: The programmer's assistant: conversational interaction with a large language model for software development. In: Proceedings of the 28th International Conference on Intelligent User Interfaces, pp. 491–514. ACM (2023). https://doi.org/10.1145/3581641.3584037
28. Smith, M.J., Sala, C., Kanter, J.M., Veeramachaneni, K.: The machine learning bazaar: harnessing the ML ecosystem for effective system development. In: Proceedings of the 2020 ACM SIGMOD International Conference on Management of Data, pp. 785–800. SIGMOD '20, Association for Computing Machinery (2020). https://doi.org/10.1145/3318464.3386146

29. Snoek, J., Larochelle, H., Adams, R.P.: Practical bayesian optimization of machine learning algorithms. In: Advances in Neural Information Processing Systems, vol. 25. Curran Associates, Inc. (2012)
30. Spitzer, P., Kühl, N., Goutier, M.: Training novices: the role of human-AI collaboration and knowledge transfer. In: Workshop on Human-Machine Collaboration and Teaming (HM-CaT 2022), The 39th International Conference on Machine Learning (2022). arXiv. https://doi.org/10.48550/ARXIV.2207.00497
31. Tayebi, A., et al.: Large language models streamline automated machine learning for clinical studies. Nat. Commun. **15**(1), 1603 (2024). https://doi.org/10.1038/s41467-024-45879-8
32. Thys, J., Vanacken, D., Rovelo Ruiz, G.: Improving AI text classification: a cascaded approach. In: 3rd Workshop on Engineering Interactive Systems Embedding AI Technologies at EICS (2025). http://hdl.handle.net/1942/46328
33. Vanbrabant, S., Eerlings, G., Rovelo Ruiz, G.A., Vanacken, D.: ECHO: enhancing conversational explainable AI through tool-augmented language models. Proc. ACM Hum.-Comput. Interact. **9**(4), EICS014:1–EICS014:33 (2025). https://doi.org/10.1145/3734191
34. Wertsch, J.V.: From social interaction to higher psychological processes a clarification and application of Vygotsky's theory. Hum. Dev. **22**(1), 1–22 (1979). https://doi.org/10.1159/000272425
35. Xiao, Z., et al.: Tell me about yourself: using an AI-powered chatbot to conduct conversational surveys with open-ended questions. ACM Trans. Comput.-Hum. Interact. **27**(3), 15:1–15:37 (2020). https://doi.org/10.1145/3381804
36. Xin, D., Wu, E.Y., Lee, D.J.L., Salehi, N., Parameswaran, A.: Whither AutoML? understanding the role of automation in machine learning workflows. In: Proceedings of the 2021 CHI Conference on Human Factors in Computing Systems, pp. 1–16. CHI '21, Association for Computing Machinery (2021). https://doi.org/10.1145/3411764.3445306
37. Yao, J., Zhang, L., Huang, J.: Evaluation of large language model-driven AutoML in data and model management from human-centered perspective. Front. Artif. Intell. **8** (2025). https://doi.org/10.3389/frai.2025.1590105
38. You, J., Park, D., Song, J.Y., Suh, B.: A labeling task design for supporting recent algorithmic needs. In: 2022 IEEE International Conference on Big Data (Big Data), pp. 2689–2698. IEEE (2022). https://doi.org/10.1109/bigdata55660.2022.10020415

MATCH: Engineering Transparent and Controllable Conversational XAI Systems Through Composable Building Blocks

Sebe Vanbrabant$^{(\boxtimes)}$, Gustavo Rovelo Ruiz , and Davy Vanacken

Hasselt University - Flanders Make, Digital Future Lab, Diepenbeek, Belgium
{sebe.vanbrabant,gustavo.roveloruiz,davy.vanacken}@uhasselt.be

Abstract. While the increased integration of AI technologies into interactive systems enables them to solve an increasing number of tasks, the black-box problem of AI models continues to spread throughout the interactive system as a whole. Explainable AI (XAI) techniques can make AI models more accessible by employing post-hoc methods or transitioning to inherently interpretable models. While this makes individual AI models clearer, the overarching system architecture remains opaque. This challenge not only pertains to standard XAI techniques but also to human examination and conversational XAI approaches that need access to model internals to interpret them correctly and completely. To this end, we propose conceptually representing such interactive systems as sequences of *structural building blocks*. These include the AI models themselves, as well as control mechanisms grounded in literature. The structural building blocks can then be explained through complementary *explanatory building blocks*, such as established XAI techniques like LIME and SHAP. The flow and APIs of the structural building blocks form an unambiguous overview of the underlying system, serving as a communication basis for both human and automated agents, thus aligning human and machine interpretability of the embedded AI models. In this paper, we present our flow-based approach and a selection of building blocks as MATCH: a framework for engineering Multi-agent Transparent and Controllable Human-Centered systems. This research contributes to the field of (conversational) XAI by facilitating the integration of interpretability into existing interactive systems.

Keywords: Intelligibility · Explainable AI · Large Language Models · Conversational XAI · Transparency · Control · User Interfaces

1 Introduction

The growing complexity of AI models, from interpretable decision trees to opaque Deep Neural Networks (DNNs) and Large Language Models (LLMs), has led to a decline in their transparency [3,29]. These models are often black boxes, producing results without explanations, justifications, or indications of uncertainties [9]. The field of eXplainable AI (XAI) addresses these challenges by complementing

C. Fayollas et al. (Eds.): EICS 2025, LNCS 16511, pp. 123–140, 2026.
https://doi.org/10.1007/978-3-032-26051-2_10

AI predictions with explanations [12,26,51]. Machine learning (ML) workflows can be made more transparent in two main ways: either by using white-box models that offer inherent interpretability, like decision trees, or by leveraging post-hoc explanations (e.g., LIME [40] and SHAP [28]) to try to explain the internal workings of black-box models, such as neural networks.

While established methods can explain the behavior of individual models, they often fall short when it comes to explaining the broader interactive systems in which those models operate. This is a significant limitation, as the quality of users' mental models of those systems directly influences how effectively they can interact with them [21]. We see similar challenges with LLMs, which require careful prompting and the right (amount of) information to respond accurately and avoid hallucinations [16]. To address the challenges of understanding the broader interactive systems that embed AI technologies, it is important to rethink how we approach the engineering of such interactive systems [7].

In the context of engineering interactive systems that incorporate LLMs, recent approaches like Tool-Augmented Language Models (TALMs) [38] and the Model Context Protocol (MCP) [2] enable LLMs to interface with external tools, supporting their integration into such systems. Furthermore, popular pipeline-based approaches like scikit-learn's pipelines and LangChain's chains provide structured ways to compose workflows of AI models and various utilities. However, these approaches remain agnostic to the user interface (UI) of the broader interactive system, which is often a conventional graphical user interface (GUI), resulting in only loosely connected conversational agents and GUIs, and thus interactive systems that fail to capitalize on the full potential of combining different UI paradigms [36,52].

In this paper, we aim to take an essential step towards engineering interactive systems with a deeper integration of conversational agents and GUIs, with the specific goal of aligning human and machine interpretations of complex systems that embed AI technologies. To this end, we introduce MATCH, a framework for engineering Multi-agent Transparent and Controllable Human-Centered systems. MATCH makes AI workflows accessible and controllable for agents (including both users and automated agents, such as LLM-based agents) through **structural** and **explanatory building blocks**. For instance, an AI model (a structural building block) can be explained using techniques like LIME [40], SHAP [28], and WhatIf [50] (explanatory building blocks). Additionally, the AI pipeline can be further controlled through structural building blocks that override decisions [18] or provide per-instance feedback [20]. We identify a selection of explanatory building blocks that address common user questions in interactive systems [23–25] and structural building blocks that support direct control of AI behavior. We incorporate these building blocks in a common pipeline-based approach, which we extend into the interactive layer by exposing individual blocks as a common knowledge base. This enables both human and automated agents to audit system behavior, which is critical for engineering responsible AI systems that are both safe and trustworthy [42]. In doing so, we address the common knowledge base highlighted by Ziegler [52].

Numerous approaches to conversational XAI exist [10,13,22,30,35,41,45]. This work centers on the engineering challenges of seamlessly integrating such approaches into existing AI workflows, combining them with GUI elements and visual explanations in a non-invasive way. To ensure the generalizability of our work, we avoid imposing strict assumptions on the structure of AI-related code, which may run in diverse environments, from experimental notebooks to systems in production. Furthermore, unlike existing approaches that typically offer explanations and insights about model behavior, our block-based framework also enables control over model behavior and execution flow. Importantly, our goal is not to replace visual XAI methods with a purely conversational approach, but to integrate both paradigms side by side. This work elaborates on the preliminary research by Vanbrabant et al. [46], with the following engineering contributions:

- **C1**: A conceptual 5-layer architecture that defines how structural and explanatory building blocks can be organized to engineer transparent and controllable interactive AI systems. The architecture provides an auditable interface accessible to multiple agents, including human and automated agents.
- **C2**: An implementation of this architecture in the form of MATCH, a modular framework that exposes developer-written AI code as composable building blocks and connects them to conversational and GUI-based interfaces.
- **C3**: A catalog of structural and explanatory building blocks, grounded in prior work, that systematically capture common forms of control and explanation in interactive AI systems.

This paper first gives an overview of relevant concepts and approaches from related work in Sect. 2. In Sect. 3, we link these together in MATCH, and we provide a catalog of structural and explanatory building blocks in Sect. 4. Finally, we reflect on our work and future directions in Sect. 5.

2   Related Work

This section explores approaches for structuring and interacting with AI systems. We examine pipeline-based and neuro-symbolic AI methods, as well as flow-based interfaces that explicate the structure and training of AI models as an inspiration for our work.

2.1   Pipelines-Based and Neuro-Symbolic AI

In a typical AI workflow, two major stages can be identified: a data-oriented stage, involving data preparation, and a model-oriented stage, involving model (re)training and deployment [1]. For supervised ML, the workflow results in a model that can predict new outputs from new inputs by leveraging its internal learning process. We can thus conceptually view a trained model as a pipeline that transforms inputs into outputs through a model. Pipelines can be chained to make system behavior more advanced and fit for a task. This pipeline-based view can be expanded to include multiple utilities. For instance, scikit-learn[1] supports pipelines to compose transformers for data preprocessing and predictors for ML models. Similarly, the LangChain framework[2] is built around chaining LLMs with various utilities, such as input prompts and structured output formatters.

Pipeline-based approaches facilitate integrating both symbolic and neural methods into a larger system. *Symbolic AI*, such as decision rules, excels at structured reasoning and provides inherent explainability and interpretability [48]. However, symbolic approaches are less trainable and more error-prone in unfamiliar situations. In contrast, *connectionist* techniques like neural networks excel at training by discovering and learning patterns from data, yet require large datasets for effective training and act as black boxes regarding explainability. *Neuro-symbolic AI (NSAI)* integrates neural and symbolic approaches to combine the individual strengths of trainability and interpretability, circumventing their inherent weaknesses [48]. Type 2 NSAI, as described by Kautz [17], considers connectionist models as neural module subroutines within a symbolic problem-solving system; the aforementioned TALMs are a recent example. TALMs query tools rather than generating answers directly, which is helpful for mathematical operations or to interface with external APIs. Furthermore, systems like ViperGPT [43] and Chameleon [27] use LLMs as neural subroutines within a symbolic tool usage framework for visual questioning answering.

2.2 Conversational and Flow-Based Interfaces for Explainable AI

LLMs offer interesting opportunities for XAI, as demonstrated in x-[plAIn] [32] and SHAPstories [31], which generate audience-specific summaries of XAI methods. These summaries are tailored to users' knowledge and interests, improving accessibility and decision-making. Such approaches, however, do not offer capabilities beyond those of the integrated XAI methods. ECHO [45] takes the conversational approach to XAI one step further, with a TALM using generated tools to explicate system-specific behavior, complemented with predefined tools that address various explanation types and XAI methods. Nevertheless, these approaches are purely textual, and would benefit from an extension with intelligible visual approaches. This works both ways: visualizations like those in TimberTrek [49] and AI-Spectra [8] could be complemented with conversational interfaces to help users understand and select the right models for their needs.

Several tools rely on flow-based, graph-like visualizations. For instance, Deep-Graph [15] visualizes models during development by constructing a data flow graph of the architecture from the DNN source code and automatically synchronizing it with a graph representation. DeepFlow [5], on the other hand, uses flow-based visual programming to realize a no-code approach to building neural networks by viewing models as sequences of learnable functions. Such approaches help visualize complex, internal architectures, but primarily focus on the development process rather than the model's role within the encapsulating system.

---

[1] https://scikit-learn.org/stable/modules/compose.html.
[2] https://github.com/langchain-ai/langchain.

In contrast, MATCH extends the scope of pipeline- and flow-based approaches for model construction and training by integrating explainability, controllability, and interaction within a unified framework. In doing so, MATCH addresses not only the technical aspect of composing AI workflows, but also the engineering challenge of making these workflows intelligible and controllable.

3 MATCH: A Framework for Engineering Multi-agent Transparent and Controllable Human-Centered Systems

Current explainability methods typically probe AI models by analyzing parts of the *input → model → output* pipeline. We expand this view by introducing building blocks that enable AI models to be systematically visualized and controlled within the larger, embedding AI system. Inspired by type 2 NSAI, we represent the AI workflow through *structural building blocks* (e.g., a neural network) that offer both human and automated agents a common knowledge base about the architecture. Each structural building block can then be further examined through *explanatory building blocks* (e.g., LIME/SHAP).

To illustrate our approach, we use the publicly available loan prediction dataset[3], a common use case in critical decision-making [11,13]. This section outlines our conceptual 5-layer architecture for engineering these building block-based systems, as well as the corresponding MATCH framework.

3.1 Auditable 5-Layer Architecture for Engineering Conversational XAI Systems Using Composable Building Blocks

Figure 1 shows our proposed 5-layer architecture, using an example ensemble trained on the loan approval prediction dataset. The vertical pipeline is loosely based on the XAI system by Mohseni et al. [33], modified to support structural and explanatory building blocks. We define the following layers, which are also depicted in Fig. 2:

Layer 1: Structural Building Blocks. These blocks convey the (conceptual) architecture of the interactive system. Each block is mapped to a part of the system's source code, as specified by the developer via functions and decorators (see Sect. 3.2). Since this representation is conceptual, developers can choose what (parts of) the underlying system to expose and how to structure/communicate the pipeline.

Layer 2: Interactive System API. Where the conceptual building blocks of layer 1 structure how developer-defined functions relate to structural components, layer 2 transforms each block's functions into an API that conveys their scope and purpose to the front-end UI and automated agents. To enable interactions, these methods are converted into a REST API. Section 3.2 details how the API generation process is put into practice.

---

[3] https://www.kaggle.com/datasets/altruistdelhite04/loan-prediction-problem-dataset.

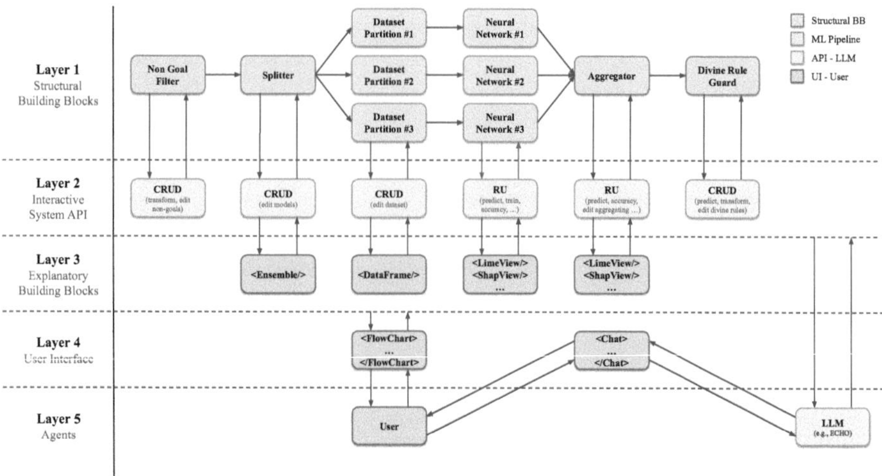

Fig. 1. An example loan approval prediction ensemble to illustrate our proposed architecture. Layer 1 shows the structural building blocks, consisting of our building blocks (in blue) integrated into the ML pipeline (in green). Layer 2 converts the structural building blocks into a callable API, usable by the explanatory building blocks in layer 3, assembled into the UI of layer 4. The LLM of layer 5 can interact with the underlying system via the API of layer 2, so that all agents operate on a common knowledge base. Agents can communicate among themselves via a chat interface. (Color figure online)

Layer 3: Explanatory Building Blocks. These interact with the structural building blocks of layer 1 through the API of layer 2. For instance, a structural building block of an AI model can have its behavior explained through LIME. Similarly, the model's training dataset can be explored through a DataFrame[4] visualization.

Layer 4: User Interface. Structural and explanatory building blocks are combined into a UI where they can be explored, interrogated, and controlled. Currently, explanatory building blocks are visualized and assembled into a coherent UI or as part of chat messages; future research directions include LLM-generated layouts that dynamically adapt to user preferences [4].

Layer 5: Agents. The final layer includes the agents that interact with the building blocks. These can be users interacting with the UI and its explanatory building blocks, or an automated LLM agent interacting with the shared API of layer 2. This API can then be integrated as a set of tools for a TALM, such as ECHO [45]. Both agents have access to the same knowledge base, allowing users, for instance, to query the LLM about system behavior.

---

[4] https://pandas.pydata.org/docs/reference/api/pandas.DataFrame.html.

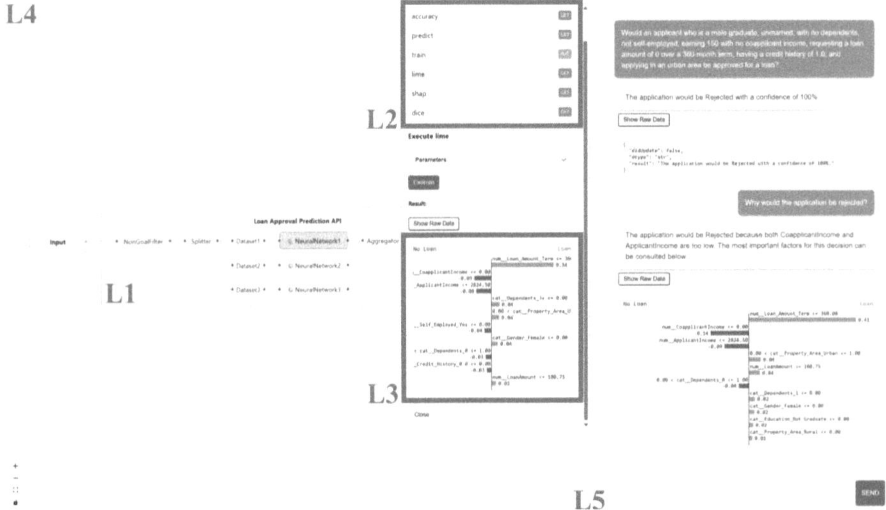

Fig. 2. Prototype of our approach, using the loan approval prediction ensemble. The UI (layer 4) shows each structural building block (layer 1) influencing predictions, exposing system behavior to both human and automated agents (layer 5) through explanatory building blocks (layer 3) and an API (layer 2), respectively. The image shows a LIME explanation for both the NeuralNetwork1 network and the total chain. Both the UI and LLM have access to the same API and are able to use the same visual components.

3.2 Implementing MATCH: Building Block Language and Multi-Agent API Integration

We designed MATCH as a lightweight integration framework that exposes existing AI workflows as composable and inspectable building blocks, as detailed in our 5-layer architecture. The central idea is to preserve the developers' existing codebase while making its internal logic auditable and manipulable through a unified interface. For instance, we assume the loan prediction AI system already exists; instead of refactoring the existing code to fit our framework, MATCH simply provides a structured way for developers to map existing functionality onto conceptual building blocks (e.g., a *neural network* block with a *predict* method). These structural building blocks clarify the methods' purpose and scope, allowing MATCH to expose the system's codebase through a generated API.

This addresses a key limitation of the ECHO framework [45], which places strict requirements on the definition of the AI models with which it can interact. In contrast, MATCH only requires developers to define which methods to expose, link them to conceptual building blocks to define their scope, and chain building blocks together to represent the overall system.

Building Block Language. In MATCH, the methods that the developer wants to expose are first linked to `BuildingBlocks` by passing them to the blocks as method lists of `Callables` (the syntax is described in EBNF in Fig. 3). Next, these building blocks are linked together in a chain that explicates the system as a pipeline. As described in Sect. 2, constructing AI workflows as pipelines is a widely adopted paradigm.

`BuildingBlocks` can be chained sequentially (denoting input-output flow) or in parallel (denoting multi-model ensemble-like approaches). In our Python implementation of MATCH, sequential chaining is facilitated by overloading the `__or__` and `__ror__` operators (pipe operator '|') on the `BuildingBlock` class, which chains two blocks together into an instance of the `Chain` class. To chain blocks in parallel, individual `Runnables` are passed as `ParallelBlock` lists. A `Runnable` can either be a `BuildingBlock`, `Chain`, or `ParallelBlock`. This approach is inspired by the LangChain framework and allows developers to express AI workflows in a concise and uniform manner.

```
Runnable        ::= BuildingBlock | Chain | ParallelBlock

BuildingBlock ::= "BuildingBlock" "(" name, "," MethodList ")"

MethodList    ::= "[" Callable* "]"

Chain           ::= Runnable "|" Runnable

ParallelBlock ::= "ParallelBlock" "(" Runnable ("," Runnable)+ ")"
```

Fig. 3. Grammar in EBNF for defining workflows consisting of building blocks in MATCH. The AI codebase is linked to individual `BuildingBlocks` as lists of `Callables` (i.e., methods). Each type of building block is a `Runnable` that can be chained sequentially via pipeline operators, or in parallel via `ParallelBlocks`.

Exposing Building Blocks to Agents. To integrate the building blocks in a front-end UI and enable conversational interactions, MATCH auto-generates a REST API of the building blocks, exposed via Flask[5]. MATCH also leverages ECHO's tool generation process for conversational XAI for the back end [45], summarized in Fig. 4.

The `API` class recursively traverses a `Runnable`, generating endpoints for each `BuildingBlock`'s methods that are marked with CRUD (i.e., Create, Read, Update, Delete) decorators. Such decorators are used in MATCH to attach additional metadata to methods. One key use of the associated CRUD metadata is to transform methods into the correct HTTP methods for the front-end-facing API and enable automatic propagation of updates through the pipeline. For example, if a dataset is modified via a method marked as `@bb_update`, any downstream building blocks (e.g., AI models whose training process depends on

---

[5] https://github.com/pallets/flask.

the dataset) will have their methods marked as `@bb_update` called automatically. Thus, updating a dataset might trigger retraining AI models further downstream automatically. Each endpoint enforces type-safe parameter parsing, executes the associated method, and serializes outputs to JSON, including metadata such as the data type and whether an update occurred. Based on this data type, the frontend and the automated agents know how to interpret a method's output.

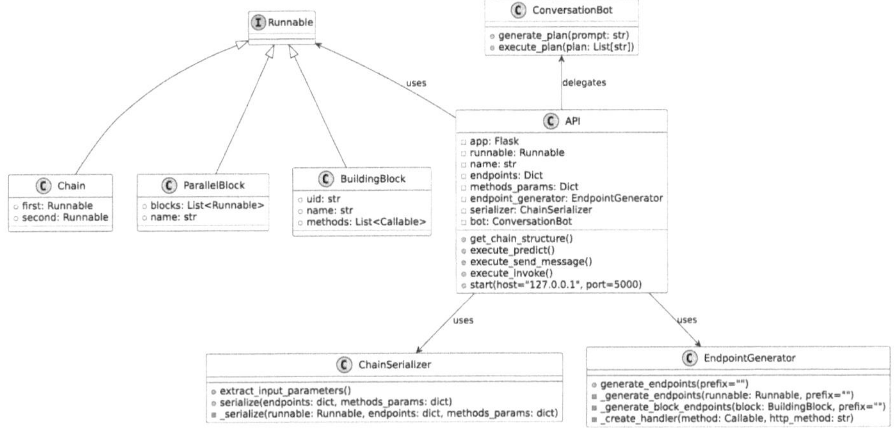

Fig. 4. Diagram of the dynamically generated API. The `API` class recursively traverses a `Runnable`, which can be a `BuildingBlock`, `Chain`, or `ParallelBlock`, to expose their methods as REST endpoints via the `EndpointGenerator` utility class. Methods annotated with CRUD or prediction decorators are automatically mapped to HTTP methods (POST, GET, PUT, DELETE). The structure of the entire `Runnable` is serialized via the `ChainSerializer` class. This workflow enables users to interact with the underlying AI workflow via a UI using the API, while the `ConversationBot` (in our case, ECHO [45]) enables conversational access via generated tools.

The API also exposes special endpoints to fetch the overall chain structure, make predictions, and initiate a conversation with ECHO, enabling both human and automated agents to query, invoke, or send messages to the chain while operating on a common knowledge base. Internally, MATCH identifies the chain's predictive method, ensuring that calls to `predict` correctly route inputs through the sequential or parallel blocks. These methods are uniquely identified by marking them as `@bb_predict`, which allows the framework to treat a chain as a functional pipeline by identifying which methods to call with which input parameters. So when a chain's `predict` method is called, inputs are passed sequentially through all `@bb_predict` methods, transforming them into a final prediction. MATCH propagates outputs of building blocks to inputs of subsequent building blocks, leaving the specific implementation and data handling of methods open to the developer. Additionally, input transformations can be triggered separately from predictions by methods marked with the `@bb_transform` decorator. This is useful, for instance, for aggregating data after retrieving it from a `ParallelBlock`, or adjusting user input to conform to intended behavior (see Sects. 4.2 and 4.1).

This design links the chain of building blocks directly to a dynamic front end, supporting both automated reasoning by automated agents such as LLMs and interactive use by human users without requiring modifications to the underlying developer-written code. While MATCH offers a novel, structured approach to interacting with AI models and pipelines by converting methods into *generated tools*, it could be further enhanced with *predefined tools*, such as the reasoning tools in Chameleon [27] or general-purpose intelligibility tools in ECHO [45].

4   Structural and Explanatory Building Blocks

This section presents an overview of possible structural and explanatory building blocks that can be integrated into the MATCH workflow. Explanatory blocks help human and automated agents interpret AI behavior, while structural blocks convey the interactive system's structure and capabilities. Structural building blocks convey a *conceptual* overview and interface of the underlying AI system, rather than imposing a strict framework. These blocks primarily use the `@bb_transform` decorator, which allows them to modify the input to a `Runnable`. By implementing CRUD operations as well, agents can further customize how the input gets modified. We illustrate our building blocks on tabular data, in accordance with the loan approval use case, although MATCH's modular workflow can be extended to other data types as well.

4.1   Structural Building Blocks for Execution Flow

We first define building blocks for visualizing and controlling (conditional) execution flows. This is particularly useful when multiple AI models are used together, such as in the context of ensemble learning [37], model multiplicity [8,49], or cascaded approaches [44], which are common in high-stakes interactive systems. Aside from the AI model and its dataset, we define a **Splitter** and **Aggregator**:

Splitter routes input data to multiple models, enabling pipelines where multiple models are used in parallel, for example, each handling different sub-tasks or data segments, or providing redundant, alternative perspectives. Users can define routing rules through a visual interface, specifying how data is directed to particular models, and optionally augmented before model training.

Aggregator aggregates the outputs from multiple models into a final decision. Users can configure aggregation strategies, such as majority voting, via a visual interface. The aggregator dashboard can display intermediate outputs and confidence levels, enhancing transparency in ensemble systems. In the context of model multiplicity, the AI-Spectra dashboard can visualize the aggregator's decision process, while Chernoff bots—robot-like adaptations of Chernoff faces [6] showing multivariate model configurations—can represent each individual model [8].

4.2   Structural Building Blocks for Control

In addition to explicating the flow between individual building blocks, we also want to make this flow, as well as the individual AI models, more controllable. For this, we primarily draw inspiration from the work on controllable AI by Kieseberg et al. [18] and map their five methods for managing control loss onto the structural building blocks in our interactive pipeline as follows:

DivineRuleGuard enforces ethical compliance by overriding harmful or unethical model outputs for non-autonomous AI systems. Users can interact with this block through a visual rule editor, which lets them define and activate conditions under which outputs must be overridden, while automated agents can generate such rules from natural language descriptions. In the loan approval context, rules could override a rejection when an applicant satisfies key financial thresholds, regardless of the model's prediction.

NonGoalFilter acts as a pre-processor, rejecting inputs that do not align with intended behaviors or intentions. Similarly to the **DivineRuleGuard**, users can interact with this block through a visual rule editor. In the loan approval context, this could be used to reject inputs where essential fields are missing or fall outside plausible ranges (e.g., negative income), preventing invalid predictions from being generated.

ShutdownTrigger functions as an emergency stop to disable autonomous AI systems at any point. While crucial for autonomous systems, it is less relevant for interactive human-in-the-loop systems where a human makes the final decision, such as our loan approval use case.

IntentionalBiasInjector strategically influences decision-making by embedding corrective signals to guide the model toward preferred outcomes. Users can re-label undesirable predictions and submit these corrections to influence future outputs. For neural networks, this can be a lightweight adaptation approach inspired by LoRA and FiLM [14,39], where only selected layers of the neural network are updated or modulated. This allows efficient personalization and customization of AI behavior without retraining the full model.

LogicBomb operates as a self-monitoring fail-safe, resetting or shutting down the AI if it ever attempts to produce an outcome that breaches ethical or operational boundaries. **LogicBomb** combines elements of **NonGoalFilter** and **ShutdownTrigger**, automatically resetting or shutting down the system based on triggered rules. This component can also monitor explanations, for instance, to guard for predictions when an explanation reveals it was mainly based on protected attributes like gender.

These control-oriented building blocks also serve a broader purpose: they represent mechanisms to calibrate the equilibrium between automated decisions and human oversight. For instance, the **DivineRuleGuard** allows human stakeholders to override automated decisions in ethically sensitive cases, ensuring that automation supports rather than replaces human judgment. Similarly, the **ShutdownTrigger** ensures humans remain in control of the system by providing an emergency stop to halt automated decisions.

4.3   Explanatory Building Blocks for AI Model Interpretability

We organize our explanatory building blocks around the commonly used explanation types of *Why, Why-not, What-if, How-to, When,* and *What-else* [25,33,45], which can be invoked straightforwardly in a conversational setting. For each type of explanation, we illustrate how established XAI methods can help in the loan approval context, and we give examples of how combining explanatory and structural building blocks can lead to more powerful systems.

Causal Explanations (Why and Why-not) describe the reasoning behind how specific inputs lead to certain outputs. For black-box models, such explanations can be provided through explanatory LIME [40] and SHAP [28] blocks, which use feature importance to address *Why* and *Why-not* questions [25]. In our loan approval example, shown in Fig. 2, consider a user wondering why their application was denied: a *Why-not*-explanation would highlight the features that most contributed to their prediction. The explanation of the system pipeline can then be cross-referenced by inspecting the individual structural building blocks (e.g., AI models or rules/filters) as they might further impact the decision or the explanations of each individual model in a multi-model setup.

Transfactual Explanations (What-if) allow users to simulate how the system responds to user-defined input values [25]. For *What-if* questions, explanatory building blocks can display the input parameters of the *predict* method and a corresponding output value, following the approaches of both He et al. [13] and Vanbrabant et al. [45]. In the loan approval example, *What-if*-explanations enable users to explore hypothetical scenarios that might lead to a more desirable outcome. They can, for instance, test alternative input values based on the features highlighted by *Why*-explanations. This can, in turn, be cross-referenced with structural building blocks that might impose further constraints on decision-making. Here, an automated agent can reason over all values and building blocks to provide holistic insights via MATCH's API.

Counterfactual Explanations (How-to) specify what changes to the inputs would yield a different output. While various methods exist [47], we adopt the DiCE framework [34], which generates diverse and plausible counterfactuals using a model-agnostic approach that supports user-defined constraints and multiple target outcomes. We visualize counterfactuals by showing original inputs and modified inputs side by side, highlighting the changes that altered the prediction. *How-to*-explanations automatically find out how users could change their loan approval status. Whereas established XAI techniques can generate counterfactuals for individual models, the API exposed by MATCH enables users and automated agents to further reason about the logic in other structural building blocks to obtain system-level counterfactuals that reflect the behavior of the entire interactive pipeline.

Generalized Prediction Conditions (When) reveal the typical conditions under which a prediction is made. Lim et al. [25] use *prototypes* to represent inputs that reliably lead to a given outcome, and *criticisms* to highlight atypical cases, helping users understand the generalizability and limits of the model's behavior. We obtain prototypes and criticisms via the MMD-critic method of Kim et al. [19], and display them in a table. When working with interpretable models like decision trees, their internal structure can be shown instead [45]. In the loan approval context, *When*-explanations help users identify and understand the typical profiles that lead to approval or rejection, such as loans being approved above a certain income threshold.

Explanations by Example (What-else) look for samples in a model's training dataset that are similar to a given input in the model representation space that produce the same or similar outputs [33]. We display *Examples* in a ranked list, ordered by similarity, with key features and prediction scores shown per instance. For loan applications, *What-else*-explanations show similar historical cases, revealing to users how comparable applicants were treated by the system. This allows them to understand whether their profile typically leads to approval or rejection, and identify the key differentiating factors among similar applications.

Integrating explanatory building blocks into MATCH offers two key advantages over standalone use. First, as discussed in Sect. 1, established XAI methods often fall short when it comes to explaining the broader interactive systems in which AI models are embedded. With MATCH, these explanatory building blocks can also be applied at the system level (which is itself also a predictive `Runnable`), as shown in Fig. 2. Second, once AI models are integrated into interactive systems, both the models and their accompanying XAI methods often become inaccessible. In contrast, MATCH's conceptual representation of complex systems keeps the embedded AI technologies accessible and interrogable to users and automated agents alike.

5   Conclusions

Existing XAI techniques often fail to scale once integrated into larger workflows and complex interactive systems. As a result, challenges regarding escalating complexity and lack of transparency persist, not only at the level of the end user but also from an engineering perspective. We addressed this with MATCH, a framework for engineering Multi-agent Transparent and Controllable Human-Centered systems. MATCH enables developers to expose their codebase through structured pipelines by chaining simple building blocks (**C2**). These structural and explanatory building blocks are integrated through an interactive systems API, situated in a conceptual 5-layer architecture. Users and automated agents such as LLMs can interact with the interactive system through GUI components or the API, respectively (**C1**). By explicitly communicating the pipeline of the interactive system, our approach provides both users and automated agents with a concrete mental model of the underlying architecture and enables them to unambiguously reason about complex pipelines.

Our catalog of building blocks is grounded in the literature: structural building blocks convey the structure of the interactive system, while explanatory building blocks offer explanatory insights into the structural components (**C3**). Through standardized CRUD operations, both users and automated agents can interrogate and control individual components via the GUI or via tool-calling in the context of TALMs. This combination of a transparent structure, explanatory views, and interactive control supports the engineering of multi-agent interactive systems that are both more transparent and more controllable compared to current pipeline-based approaches. Although we have set our current focus on tabular data, MATCH's modular approach can be extended to other data types as well.

MATCH can benefit multiple stakeholders in distinct ways. Firstly, developers gain a minimally invasive mechanism to expose system functionality as composable building blocks. Secondly, designers can reason about the underlying system architecture through visualized workflows, facilitating the integration of human-centered design practices. Thirdly, end users can interact with explanations and control mechanisms, enabling them to interrogate and influence system behavior directly. By supporting multiple perspectives, MATCH contributes to a more comprehensive engineering approach to embedding AI technologies in interactive systems. Furthermore, the modular nature of MATCH makes it straightforward to add additional building blocks or conceptual layers that can further target diverse stakeholders.

Acknowledgments. This work was funded by the Special Research Fund (BOF) of Hasselt University, BOF23OWB31.

Disclosure of Interests. The authors have no competing interests to declare that are relevant to the content of this article.

References

1. Amershi, S., Begel, A., Bird, C., DeLine, R., Gall, H., Kamar, E., et al.: Software engineering for machine learning: a case study. In: 2019 IEEE/ACM 41st International Conference on Software Engineering: Software Engineering in Practice (ICSE-SEIP), pp. 291–300 (2019). https://doi.org/10.1109/ICSE-SEIP.2019.00042
2. Anthropic: Introducing the Model Context Protocol (2024). https://www.anthropic.com/news/model-context-protocol
3. Barredo Arrieta, A., Díaz-Rodríguez, N., Del Ser, J., Bennetot, A., Tabik, S., Barbado, A., et al.: Explainable Artificial Intelligence (XAI): concepts, taxonomies, opportunities and challenges toward responsible AI. Inform. Fusion **58**, 82–115 (2020). https://doi.org/10.1016/j.inffus.2019.12.012
4. Brie, P., Burny, N., Sluÿters, A., Vanderdonckt, J.: Evaluating a large language model on searching for GUI layouts. Proc. ACM Hum.-Comput. Interact. **7**(EICS) (2023). https://doi.org/10.1145/3593230
5. Calò, T., De Russis, L.: DeepFlow: a flow-based visual programming tool for deep learning development. In: Proceedings of the 30th International Conference on Intelligent User Interfaces, IUI 025, pp. 504–518. Association for Computing Machinery, New York (2025). https://doi.org/10.1145/3708359.3712109

6. Chernoff, H.: The use of faces to represent points in K-dimensional space Graphically. J. Am. Statist. Associa. **68**(342), 361–368 (1973), http://www.jstor.org/stable/2284077

7. Dix, A., Mayer, S., Palanque, P., Panizzi, E., Spano, L.D.: Engineering interactive systems embedding AI technologies. In: Companion Proceedings of the 2023 ACM SIGCHI Symposium on Engineering Interactive Computing Systems, EICS 2023 Companion, pp. 90–92. Association for Computing Machinery, New York (2023). https://doi.org/10.1145/3596454.3597195

8. Eerlings, G., Vanbrabant, S., Liesenborgs, J., Rovelo Ruiz, G., Vanacken, D., Luyten, K.: AI-Spectra: a visual dashboard for model multiplicity to enhance informed and transparent decision-making. In: Zaina, L., Campos, J.C., Spano, D., Luyten, K., Palanque, P., van der Veer, G., et al. (eds.) Engineering Interactive Computer Systems. EICS 2024 International Workshops, pp. 55–73. Springer Nature Switzerland, Cham (2025). https://doi.org/10.1007/978-3-031-91760-8_5

9. von Eschenbach, W.J.: Transparency and the black box problem: why we do not trust AI. Philos. Technol. **34**(4), 1607–1622 (2021). https://doi.org/10.1007/s13347-021-00477-0

10. Garofalo, M., Fantini, A., Pellugrini, R., Pilato, G., Villari, M., Giannotti, F.: Conversational XAI: formalizing its basic design principles. In: Meo, R., Silvestri, F. (eds.) Machine Learning and Principles and Practice of Knowledge Discovery in Databases, pp. 295–309. Springer Nature Switzerland, Cham (2025)

11. Green, B., Chen, Y.: The principles and limits of algorithm-in-the-loop decision making. Proc. ACM Hum.-Comput. Interact. **3**(CSCW) (2019). https://doi.org/10.1145/3359152

12. Gunning, D., Stefik, M., Choi, J., Miller, T., Stumpf, S., Yang, G.Z.: XAI—explainable artificial intelligence. Sci. Robot. **4**(37), eaay7120 (2019). https://doi.org/10.1126/scirobotics.aay7120

13. He, G., Aishwarya, N., Gadiraju, U.: Is Conversational XAI all you need? human-AI decision making with a conversational XAI assistant. In: Proceedings of the 30th International Conference on Intelligent User Interfaces, IUI 2025, pp. 907–924. Association for Computing Machinery, New York(2025). https://doi.org/10.1145/3708359.3712133

14. Hu, E.J., Shen, Y., Wallis, P., Allen-Zhu, Z., Li, Y., Wang, S., et al.: LoRA: Low-Rank Adaptation of Large Language Models (2021), https://arxiv.org/abs/2106.09685

15. Hu, Q., Ma, L., Zhao, J.: DeepGraph: a pycharm tool for visualizing and understanding deep learning models. In: 2018 25th Asia-Pacific Software Engineering Conference (APSEC), pp. 628–632 (2018). https://doi.org/10.1109/APSEC.2018.00079

16. Huang, L., Yu, W., Ma, W., Zhong, W., Feng, Z., Wang, H., et al.: A survey on hallucination in large language models: principles, taxonomy, challenges, and open questions. ACM Trans. Inf. Syst. **43**(2) (2025). https://doi.org/10.1145/3703155

17. Kautz, H.: The Third AI Summer: AAAI Robert S Engelmore Memorial Lecture. AI Mag. **43**(1), 105–125 (2022). https://doi.org/10.1002/aaai.12036

18. Kieseberg, P., Weippl, E., Tjoa, A.M., Cabitza, F., Campagner, A., Holzinger, A.: Controllable AI - an alternative to trustworthiness in complex AI systems? In: Holzinger, A., Kieseberg, P., Cabitza, F., Campagner, A., Tjoa, A.M., Weippl, E. (eds.) Machine Learning and Knowledge Extraction. pp. 1–12. Springer Nature Switzerland, Cham (2023). https://doi.org/10.1007/978-3-031-40837-3_1

19. Kim, B., Khanna, R., Koyejo, O.O.: Examples are not enough, learn to criticize! Criticism for Interpretability. In: Lee, D., Sugiyama, M., Luxburg, U., Guyon, I., Garnett, R. (eds.) Advances in Neural Information Processing Systems, vol. 29. Curran Associates, Inc. (2016), https://proceedings.neurips.cc/paper_files/paper/2016/file/5680522b8e2bb01943234bce7bf84534-Paper.pdf
20. Kulesza, T., Burnett, M., Wong, W.K., Stumpf, S.: Principles of explanatory debugging to personalize interactive machine learning. In: Proceedings of the 20th International Conference on Intelligent User Interfaces, IUI 2015, pp. 126–137. Association for Computing Machinery, New York (2015).https://doi.org/10.1145/2678025.2701399
21. Kulesza, T., Stumpf, S., Burnett, M., Kwan, I.: Tell me more? the effects of mental model soundness on personalizing an intelligent agent. In: Proceedings of the SIGCHI Conference on Human Factors in Computing Systems, CHI 2012, pp. 1–10. Association for Computing Machinery, New York (2012). https://doi.org/10.1145/2207676.2207678
22. Kuźba, M., Biecek, P.: What would you ask the machine learning model? identification of user needs for model explanations based on human-model conversations. In: Koprinska, I., et al. (eds.) ECML PKDD 2020. CCIS, vol. 1323, pp. 447–459. Springer, Cham (2020). https://doi.org/10.1007/978-3-030-65965-3_30
23. Lim, B.Y., Dey, A.K.: Assessing demand for intelligibility in context-aware applications. In: Proceedings of the 11th International Conference on Ubiquitous Computing, UbiComp 2009, pp. 195–204. Association for Computing Machinery, New York (2009). https://doi.org/10.1145/1620545.1620576
24. Lim, B.Y., Dey, A.K.: Toolkit to support intelligibility in context-aware applications. In: Proceedings of the 12th ACM International Conference on Ubiquitous Computing, UbiComp 2010, pp. 13–22. Association for Computing Machinery, New York (2010). https://doi.org/10.1145/1864349.1864353
25. Lim, B.Y., Yang, Q., Abdul, A.M., Wang, D.: Why these explanations? selecting intelligibility types for explanation goals. In: Explainable Smart Systems Workshop at IUI (2019). https://ceur-ws.org/Vol-2327/IUI19WS-ExSS2019-20.pdf
26. Longo, L., Brcic, M., Cabitza, F., Choi, J., Confalonieri, R., Ser, J.D., et al.: Explainable artificial intelligence (XAI) 2.0: a manifesto of open challenges and interdisciplinary research directions. Inform. Fusion **106**, 102301 (2024). https://doi.org/10.1016/j.inffus.2024.102301
27. Lu, P., Peng, B., Cheng, H., Galley, M., Chang, K.W., Wu, Y.N., et al.: Chameleon: plug-and-Play Compositional Reasoning with Large Language20Models. In: Proceedings of the 37th International Conference on Neural Information Processing Systems, NIPS '23, Curran Associates Inc., Red Hook, NY, USA (2023)
28. Lundberg, S.M., Lee, S.I.: A unified approach to interpreting model predictions. In: Proceedings of the 31st International Conference on Neural Information Processing Systems, NIPS 2017, pp. 4768–4777. Curran Associates Inc., Red Hook, NY, USA (2017)
29. Luo, H., Specia, L.: From Understanding to Utilization: A Survey on Explainability for Large Language Models (2024). https://arxiv.org/abs/2401.12874
30. Malandri, L., Mercorio, F., Mezzanzanica, M., Nobani, N.: ConvXAI: a System for Multimodal Interaction with Any Black-box Explainer. Cogn. Comput. **15**(2), 613–644 (2023). https://doi.org/10.1007/s12559-022-10067-7
31. Martens, D., Hinns, J., Dams, C., Vergouwen, M., Evgeniou, T.: Tell me a story! narrative-driven XAI with large language models. Decis. Support Syst. **191**, 114402 (2025). https://doi.org/10.1016/j.dss.2025.114402

32. Mavrepis, P., Makridis, G., Fatouros, G., Koukos, V., Separdani, M.M., Kyriazis, D.: XAI for All: Can Large Language Models Simplify Explainable AI? (2024). https://arxiv.org/abs/2401.13110
33. Mohseni, S., Zarei, N., Ragan, E.D.: A multidisciplinary survey and framework for design and evaluation of explainable AI systems. ACM Trans. Interact. Intell. Syst. **11**(3–4) (2021). https://doi.org/10.1145/3387166
34. Mothilal, R.K., Sharma, A., Tan, C.: Explaining machine learning classifiers through diverse counterfactual explanations. In: Proceedings of the 2020 Conference on Fairness, Accountability, and Transparency, FAT* 2020, pp. 607–617. Association for Computing Machinery, New York (2020). https://doi.org/10.1145/3351095.3372850
35. tterer, J., Seifert, C.: From black boxes to conversations: incorporating XAI in a conversational agent. In: Longo, L. (ed.) Explainable Artificial Intelligence, pp. 71–96. Springer Nature Switzerland, Cham (2023) . https://doi.org/10.1007/978-3-031-44070-0_4
36. Nielsen, J.: AI: First New UI Paradigm in 60 Years (2023). https://www.nngroup.com/articles/ai-paradigm/
37. Opitz, D., Maclin, R.: Popular ensemble methods: an empirical study. J. Artif. Int. Res. **11**(1), 169–198 (1999)
38. Parisi, A., Zhao, Y., Fiedel, N.: TALM: Tool Augmented Language Models (2022). https://arxiv.org/abs/2205.12255
39. Perez, E., Strub, F., de Vries, H., Dumoulin, V., Courville, A.: FiLM: visual reasoning with a general conditioning layer. In: Proceedings of the Thirty-Second AAAI Conference on Artificial Intelligence and Thirtieth Innovative Applications of Artificial Intelligence Conference and Eighth AAAI Symposium on Educational Advances in Artificial Intelligence, AAAI 2018/IAAI 2018/EAAI 2018, AAAI Press (2018)
40. Ribeiro, M.T., Singh, S., Guestrin, C.: "Why should i trust you?": explaining the predictions of any classifier. In: Proceedings of the 22nd ACM SIGKDD International Conference on Knowledge Discovery and Data Mining, KDD 2016, pp. 1135–1144. Association for Computing Machinery, New York (2016). https://doi.org/10.1145/2939672.2939778
41. Slack, D., Krishna, S., Lakkaraju, H., Singh, S.: Explaining machine learning models with interactive natural language conversations using TalkToModel. Nat. Mach. Intell. **5**(8), 873–883 (2023). https://doi.org/10.1038/s42256-023-00692-8
42. Stumpf, S., et al.: Engineering safe and trustworthy AI: the participatory harm auditing workbenches and methodologies (PHAWM) project. In: Workshop on Engineering Interactive Systems Embedding AI Technologies at EICS (2025)
43. Surís, D., Menon, S., Vondrick, C.: ViperGPT: visual inference via python execution for reasoning. In: 2023 IEEE/CVF International Conference on Computer Vision (ICCV), pp. 11854–11864 (2023). https://doi.org/10.1109/ICCV51070.2023.01092
44. Thys, J., Vanacken, D., Rovelo Ruiz, G.: Improving AI text classification: a cascaded approach. In: Workshop on Engineering Interactive Systems Embedding AI Technologies at EICS (2025)
45. Vanbrabant, S., Eerlings, G., Rovelo Ruiz, G., Vanacken, D.: ECHO: enhancing conversational explainable AI through tool-augmented language models. Proc. ACM Hum.-Comput. Interact. **9**(4) (2025). https://doi.org/10.1145/3734191

46. Vanbrabant, S., Rovelo Ruiz, G., Vanacken, D.: Composable building blocks for controllable and transparent interactive AI systems. In: Workshop on Engineering Interactive Systems Embedding AI Technologies at EICS (2025). https://arxiv.org/abs/2506.02262
47. Wachter, S., Mittelstadt, B., Russell, C.: Counterfactual Explanations without Opening the Black Box: Automated Decisions and the GDPR (2018). https://arxiv.org/abs/1711.00399
48. Wang, W., Yang, Y., Wu, F.: Towards data-and knowledge-driven ai: a survey on neuro-symbolic computing. IEEE Trans. Pattern Anal. Mach. Intell. **47**(2), 878–899 (2025). https://doi.org/10.1109/TPAMI.2024.3483273
49. Wang, Z.J., Zhong, C., Xin, R., Takagi, T., Chen, Z., Chau, D.H., et al.: TimberTrek: exploring and curating sparse decision trees with interactive visualization. In: 2022 IEEE Visualization and Visual Analytics (VIS), pp. 60–64 (2022). https://doi.org/10.1109/VIS54862.2022.00021
50. Wexler, J., Pushkarna, M., Bolukbasi, T., Wattenberg, M., Viégas, F., Wilson, J.: The what-if tool: interactive probing of machine learning models. IEEE Trans. Visual Comput. Graphics **26**(1), 56–65 (2020). https://doi.org/10.1109/TVCG.2019.2934619
51. Xu, X., Yu, A., Jonker, T.R., Todi, K., Lu, F., Qian, X., et al.: XAIR: a framework of explainable ai in augmented reality. In: Proceedings of the 2023 CHI Conference on Human Factors in Computing Systems, CHI 2023, Association for Computing Machinery, New York (2023). https://doi.org/10.1145/3544548.3581500
52. Ziegler, J.: Challenges in integrating conversational AI and GUI-based applications. In: Zaina, L., Campos, J.C., Spano, D., Luyten, K., Palanque, P., van der Veer, G., et al. (eds.) Engineering Interactive Computer Systems. EICS 2024 International Workshops, pp. 159–170. Springer Nature Switzerland, Cham (2025). https://doi.org/10.1007/978-3-031-91760-8_11

Experience 2.0 and Beyond – Engineering Cross Devices and Multiple Realities (EXDMR 2025)

Wayfinding and Cognitive Mapping in Virtual Reality in Complex Buildings Using Outdoor and Indoor Landmarks

Nils Ove Beese[1,2], Jan Spilski[1(✉)], Thomas Lachmann[1,3,4],
Jan-Hendrik Sünderkamp[5], Jan Hendrik Plümer[5], Alexander Jaksties[6],
and Kerstin Müller[6]

[1] Center for Cognitive Science, RPTU Kaiserslautern-Landau, Kaiserslautern, Germany
`nils.beese@med.uni-goettingen.de`, `{spilski,lachmann}@rptu.de`
[2] Department of Occupational, Social and Preventive Medicine, University Medical Center Göttingen, Göttingen, Germany
[3] Centro de Investigación Nebrija en Cognición, Facultad de Lenguas y Educación, Universidad Nebrija, Madrid, Spain
[4] Cognition Research Unit, KU Leuven, Leuven, Belgium
[5] Creative Technologies, Salzburg University of Applied Sciences, Salzburg, Austria
`{jan-hendrik.suenderkamp,jan.pluemer}@fh-salzburg.ac.at`
[6] CGTI Lab, HS Bielefeld, Bielefeld, Germany
`{alexander.jaksties,kerstin.mueller}@hsbi.de`

Abstract. Wayfinding is an important skill in the context of everyday life. Especially in unknown environments, spatial orientation in real life relies heavily on landmarks. The present study investigated the impact of outdoor and indoor landmarks in a wayfinding task performed in unknown complex office buildings in Virtual Reality. In order to investigate the performance measures of orientation, twenty-two participants had to find the conference room in the office buildings. The office buildings were constructed in the manner of mazes, incorporating dead ends and loops, and were constructed as two-story buildings. These buildings had no landmarks, indoor landmarks, outdoor landmarks or both types of landmarks. At the end, participants had to draw a digital sketch map to show the way to the conference room. The presence of landmarks led to more accurate sketch maps, in comparison to conditions without landmarks. While a discrepancy in accuracy was observed between the outdoor and the indoor landmark conditions in the data, this discrepancy did not attain statistical significance.

Keywords: Spatial Cognition · Virtual Reality · User Performance

1 Introduction

Spatial orientation is an essential cognitive ability that facilitates navigation in various environments, such as urban settings, medical facilities, and office buildings [1]. This ability is especially crucial in safety-critical situations, such as locating emergency

© The Author(s), under exclusive license to Springer Nature Switzerland AG 2026
C. Fayollas et al. (Eds.): EICS 2025, LNCS 16511, pp. 143–160, 2026.
https://doi.org/10.1007/978-3-032-26051-2_11

exits or efficiently navigating unfamiliar buildings [2–4]. There is evidence that indoor landmarks and outdoor landmarks increase orientation performance [4, 5]. However, there is a lack of experimental comparisons between conditions with and without visible external references, especially in complex indoor spaces [1, 6]. Outdoor landmarks may provide global reference points that complement or even enhance indoor orientation cues. Understanding how these external landmarks affect navigation inside buildings could have practical implications for architecture, safety design, and optimizing human–environment interaction [e.g., 5]. Virtual Reality (VR) offers a unique opportunity to address this research gap. VR combines ecological validity with high experimental control, enabling researchers to simulate complex indoor environments and systematically manipulate landmark availability [7]. Previous studies have shown that spatial representations acquired in VR closely resemble those in the real world, and that they can be transferred between environments [8–10]. Building upon the foundations established by Jaksties et al. [11] and Müller et al. [12], the present experimental study extends their research on the beneficial effects of landmarks on orientation in VR. However, these studies focused primarily on indoor landmarks in closed buildings without windows. Our study addresses this limitation by investigating the influence of outdoor landmarks on wayfinding inside a complex office-like building in VR. Therefore, our experimental study advances the understanding of how different types of landmarks affect spatial orientation within complex buildings.

2 Related Work

Spatial cognition is the abilitiy to use spatial information about one's surroundings to accomplish goals, such as identifying objects, using representations of the world, like maps, and navigating through the world itself [14]. Spatial orientation is one of the most prominent spatial cognition abilities. It refers to perceiving and adjusting one's location according to objects in the environment [15]. Orientation relies on spatial knowledge acquisition as well as representation. One of the most well-established models of spatial knowledge acquisition is the landmark, route and survey knowledge model, as described by Siegel and White [16] and Thorndyke and Goldin [17]. This model describes spatial knowledge as a three-stage theory. The first stage of acquisition is landmark knowledge. The theory argues that landmarks are initially extracted from any given environment due to their salience and orientation dependence [18]. It is presumed that this knowledge takes the form of perceptual images. This kind of knowledge is acquired directly through vision. Recognition of a location, and thus orientating oneself in any environment, depends heavily on landmark knowledge [17]. The second stage in Thorndyke and Goldin's model is route or procedural knowledge. Route knowledge is an integral part of navigation because it encodes the spatial relationship between points A and B via the route and actions that connect them. The third stage of the model is survey knowledge, also termed configurational knowledge. The representation of distances between and locations of objects is analogous to that of a standard map. This type of knowledge develops through repeated navigation or by learning the map of an environment [16].

While Siegel and White [16] claim that acquisition happens serially, more recent research [19–21] argues that acquisition happens in a parallel. A major point in spatial knowledge acquisition is also movement control [23]. Movement control is said to facilitate the acquisition of both landmark knowledge and route knowledge [22, 23].

Perhaps the most essential skill in navigation is wayfinding, which is also sometimes referred to as pathfinding. It describes the basic cognitive process of reaching a destination [24, 25]. According to Freundschuh [25], navigation is the combination of locomotion (i.e., moving in an environment) and wayfinding. According to Allen [26], there are three categories of wayfinding tasks: exploratory wayfinding, commute-like wayfinding and wayfinding to novel destinations, also called quest wayfinding [25]. Exploratory wayfinding describes the process of becoming familiar with an unfamiliar environment. Commute-like wayfinding involves traveling between two known points on a familiar route. The goal of quest wayfinding is to reach an unknown destination (target) from a known starting point. Moura and Bartram [27] investigated players' responses to different wayfinding cues in 3D games using a wayfinding paradigm. Moura and Bartram discovered that wayfinding cue characteristics must be adapted to the challenge. For example, clear cues and landmarks are necessary for spaces with difficult visibility, such as mazes. According to Freundschuh [25] there are several wayfinding strategies, such as: piloting, habitual locomotion, and the use of internal representations. Piloting relies heavily on landmarks: it involves travelling from one landmark to another while exploring the environment. Habitual locomotion typically develops after repeated navigation in an environment. The use of internal representations is the most sophisticated of these three strategies. As the name suggests, this strategy requires an internal representation of the environment one needs to traverse. All three strategies rely on cognitive maps and landmarks.

Landmarks play a vital role in orientating in an unknown environment. The uncertainty of an unknown environment lends itself to search for landmarks to facilitate orientation [28, 29]. Grzeschik et al. [30] showed that landmarks placed at intersections support the development of route knowledge. Similarly, Wang et al. [31] varied landmark conditions and demonstrated that participants could use landmark knowledge for guidance after only one trial in their experimental setting. Jansen-Osmann and Fuchs [20] showed that landmarks can facilitate orientation. Von Stülpnagel and Steffens [23] demonstrated that using a combination of landmarks and navigation maps might be disadvantageous compared to using only landmarks in route knowledge tasks.

Edward Tolman [32] introduced cognitive maps as mental representations of a spatial environment in 1948. Tolman demonstrated that rats could navigate a maze correctly even when the starting point was changed. Warren et al. [33] argue in their study that a graph-like representation might be a better analogy than that of a map. Weisberg and Newcombe [34] suggest that the two approaches, "cognitive map" and "cognitive graph", have common aspects, integrating both views while also highlighting differences in individuals' performances. Peer et al. [35] further argue for the existence of both cognitive maps and cognitive graphs depending on the task and the spatial information provided by the environment. Sketch maps are a reliable and valid method to measure cognitive maps or graphs in real-world environments as well as virtual environments (VE)

[36]. Previous research has shown that sketch maps can be used to measure wayfinding performance [37] and are not influenced by drawer's expertise [38].

Regarding navigation, Weisberg and colleagues [40] found that learning patterns in VEs were similar to those in the real world, though accuracy was higher in the real world. Ruddle et al. [41] compared navigation performance using a desktop VE versus an HMD VE and showed that participants could navigate faster and had a better assessment of the direct distance between two places when using the HMD VE. However, the ability to estimate the direction of different places in relation to one's own position was the same in both environments. Jansen-Osmann [42] showed that landmarks improve orientation when finding a way and that routes with landmarks are memorized faster than routes without landmarks when using a desktop VE. Grzeschik et al. [30] showed that landmarks can also improve the ability to navigate novel routes. Participants performed better in conditions with different, unique landmarks. Grzeschik et al. [30] argue that this may be because participants did not need detailed knowledge of the spatial configurations of similar landmarks in the different Landmark conditions. Ruddle et al. [43] investigated the impact of local and global landmarks on route knowledge in a virtual marketplace. Participants navigated a desktop VE four times: once without landmarks, once only global landmarks, once with only local landmarks, and once with both global and local landmarks. In their first experiment, local landmarks reduced the number of errors participants made, but global Landmarks did not. Ruddle and colleagues suggest that recall of direction at decision points may be influenced by a landmark-action pair rather than by the visual cue of the landmark itself. According to Stankiewicz and Kalia [13], there is also a possible competition for cognitive resources when different types of landmarks available. Stankiewicz and Kalia setup three experiment to test structural and object landmarks and the acquisition of landmark knowledge. For object landmarks, they used pictures placed on the hallway walls. In their first experiment, they examined how structural and object landmarks might differ in knowledge acquisition if the information content of both is equal. The structural landmarks were remembered more accurately than the object landmarks. In their second experiment, Stankiewicz and Kalia [13] investigated how the increased information content of object landmarks affects accuracy. There was no difference between the structural and object landmarks. In their third experiment, the object landmarks were arranged in a way that they were identical in adjacent hallways. This meant that the structural and object landmarks were not independent from each other. Structural landmarks were remembered more accurately than object landmarks. Müller et al. [12] developed a virtual reality (VR) software to explore connection between spatial cognition and different factors, such as landmarks, environmental complexity, movement methods, and texture. Participants had to reach a target room from the starting room as quickly as possible while traversing as few rooms as possible. The results of their study showed a positive impact of indoor landmarks. However, their setup was in a closed building without windows.

3 Hypotheses

Most prior VR studies used windowless buildings and focused on indoor landmarks and easy environments for wayfinding tasks [e.g., 4, 12, 44], and focusing on search time [45, 46]. Our study addresses these limitations by including both indoor and outdoor landmarks, using an unknown complex maze-like, two-floor office building, and measure performance in the sketch map task, based on the time-to-orientation and accuracy of the sketch map. Therefore, our research goal was to explore how landmark presence and type affect wayfinding performance and cognitive map development in a quest wayfinding task. In our study, we compared two conditions: environments with and without landmarks. We measured orientation performance using sketched cognitive maps. We hypothesize that *orientation performance in VR will be better in conditions with landmarks compared to conditions without landmarks* **(H1)**.

Hypothesis 2 focuses on the type of landmarks and their effect on orientation performance. The assumption was that outdoor landmarks can be good overall cues, because they are visible through windows in different rooms, provide an understanding of broad cardinal direction [47, 48], thereby helping with general orientation. Therefore, we hypothesize that *Orientation performance will benefit more from outdoor landmarks than from indoor landmarks* **(H2)**.

Finally, for Hypothesis 3, we compared two conditions: The outdoor landmark only condition and the combined condition, which includes both indoor and outdoor landmarks. While more cues could helpful, they could also cause interference or overload. Competition for cognitive resources and possibly split attention should result in longer processing times for participants [13, 49, 50]. Therefore, we hypothesize that *the time for orientation assessment will be better in an outdoor landmark only condition compared to a combined indoor and outdoor landmark condition* **(H3)**.

4   Methods

4.1   Participants

All participants were recruited via the university email newsletter and received financial compensation or study credits for participating. Twenty-five individuals (18 male and 7 female) between the ages of 18 and 31 ($M = 24.28$, $SD = 3.87$) participated in the user study. Three participants had to cancel due to severe symptoms of simulator sickness after the first building. Their data were omitted from the analysis due to their withdrawal. Only the data from the remaining 22 participants were analyzed. All participants reported having normal or corrected-to-normal vision and hearing. 12 participants reported previous VR experience. In order to exclude possible biases between participants with and without prior VR experience in the main analyses (hypothesis testing), U-tests were carried out. We found no statistically significant difference between participants with and without VR experience for our independent variables (*"inaccuracy of orientation"* and *"time for orientation"*, p's $> .26$). According to the Karolinska Sleepiness Scale [51], the participants were mostly rather alert ($Md = 3$, $IQR = 1$). In order to quickly identify any problems for participants and in accordance with ethical principles, we also tested simulator sickness and different workload factors, with the Simulator Sickness Questionnaire [SSQ, 52] and the NASA Task Load Index [NASA-TLX, 53]. Participants reported none or mild to moderate simulator sickness symptoms and a task load ranging from low to high (*Range M* $= 6.86$ to 15.68).

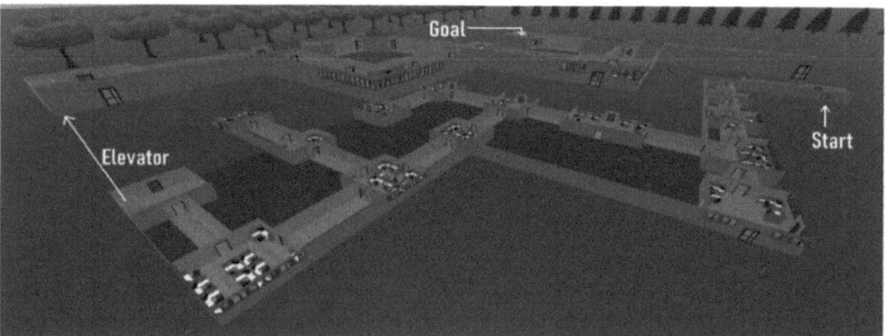

Fig. 1. Insight into a generated two-story building.

4.2 Prototype of the Virtual Environment

The environment and setup of the experiment are based on the aforementioned studies [11, 12, 43], taking their findings into account. Therefore, we developed an application that generates two-story buildings with randomized room structures (Fig. 1). We generated five different buildings. One of these buildings was designed for the tutorial and had a ground floor with five rooms. This building was used in the tutorial to familiarize players with the setting and the VR controllers. The other four buildings differed in terms of the landmark conditions. The landmarks conditions were: no landmarks; only indoor landmarks; only outdoor landmarks; and both indoor and outdoor landmarks.

The two-story buildings were designed to investigate orientation performance at different levels of the building. The ground floor of these buildings consisted of eleven rooms along the main path, which had three junctions, as well as five additional rooms. Each upper floor had seven rooms along the main path and a loop. Additionally, there was one extra junction leading to a dead end. First, the entrance room was placed. Starting from this room, the main path connecting the entrance with the elevator was generated. All of the rooms were rectangular. They looked very similar. They were equipped with office furniture, such as cubicles containing chairs, keyboards and monitors, and desks, which were arranged either close to the walls, in the middle of the room, or both. The lighting was flat with no shadows, in order to eliminate any possible effects due to the shape and size of the landmarks, particularly the outdoor landmarks.

Fig. 2. Big office room with windows on the right side and a closed door on the left side.

Generating the upper floor followed the same pattern as generating the ground floor. The main difference is that the main path began at the elevator and ended at a specific office room representing the target. After placing all the rooms, their entrances were connected through corridors. Figure 2 shows an example of a possible junction with outdoor landmarks and closed doors. Depending on the generated buildings and rooms, as well as their placement within the buildings, these larger rooms might have two, three or four open doors, as well as two, one or no closed doors or door-like windows. As shown in Fig. 2, windows and door-like windows were present in every building, regardless of the landmark condition.

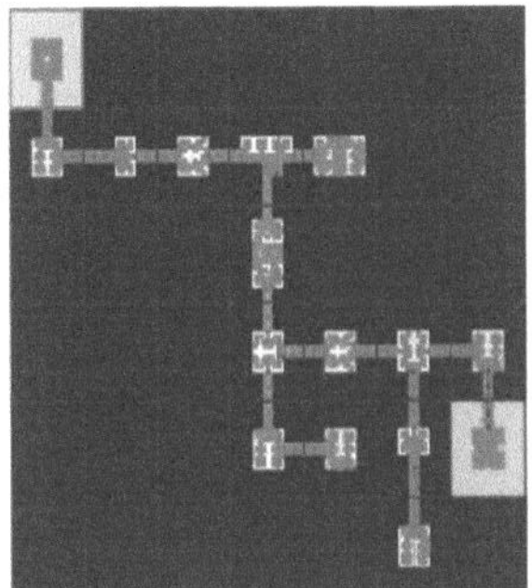 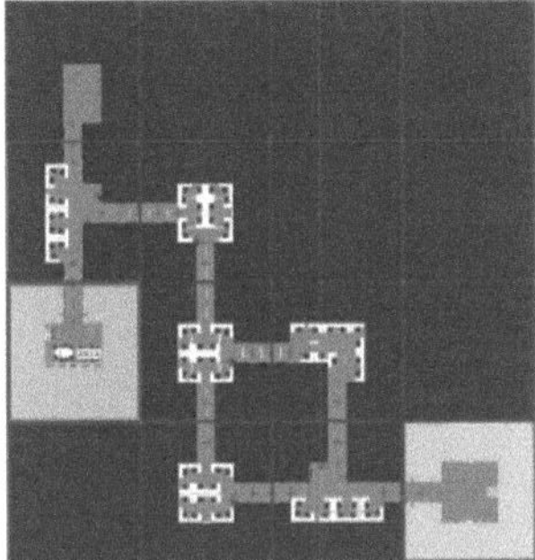

Fig. 3. Exemplary layout of the floors in a generated building. Left: The ground floor of a building. The start room is in the chunk marked green and the elevator in the chunk marked blue. Right: The upper floor of a building. The elevator is in the chunk marked blue and the target room is in the chunk marked violet.

Figure 3 shows an example of the ground and upper floors of a generated building. Outside the building, distinct landscapes were present in each cardinal direction, such as two types of trees, a mountain, and historic buildings that serve as outdoor landmarks (Fig. 4). If outdoor landmarks were not present, these were made invisible, and outside of the building was just a simple green plane mimicking grass.

A VR setup with the HTC Vive Pro Eye and Vive controllers was used to present the virtual building. The HMD has a resolution of 1440 x 1600 Pixel per eye, a field of view of 110°, and a refresh rate of 90 Hz. The player started at the entrance of the building (Fig. 5a), where the cardinal directions are indicated by letters above the doors (N, S, E, W) and an empty grid map including the cardinal directions. The objective was to locate and access the conference room (Figure 5b) on the upper floor. Due to space constraints, the participants used the arm-swinging technique [54] to move through the building. They had to press the Grip Button on both controllers and swing their arms as if walking in place. This movement form was implemented using the ArmSwinger VR Locomotion System by ElectricNightOwl [55]. Upon reaching the end of a floor, the players had to sketch the main path of that floor in the VR environment. They could do so by selecting the chunks on a grid map with their virtual hands.

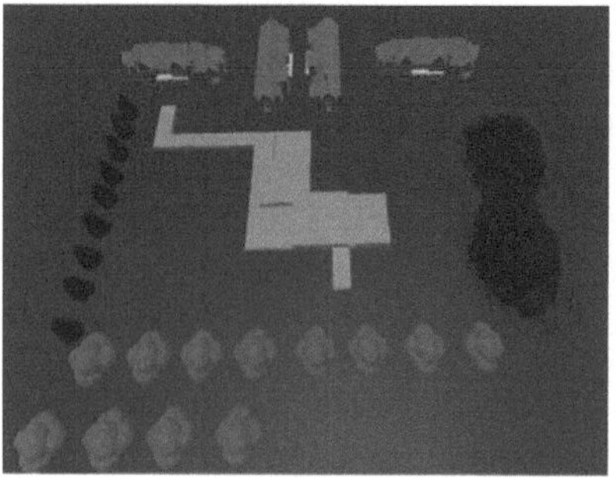 

Fig. 4. Left: Outdoor landmarks: buildings in the north, mountain in the east, green pine trees in the west and autumnal leaf trees in the south. Right: Indoor landmarks: Possible couch landmark placed at a junction point of the office.

a   b

Fig. 5 a. Start room on the ground floor. b. Goal: Target on the upper floor.

4.3   General Procedure and Outcomes

The experiment complies with the APA Code of Ethics [56], the WMA Declaration of Helsinki [57], and has been approved by the ethics committee of the University of Kaiserslautern-Landau. At the beginning, the participants received information about the experiment procedure and provided informed written consent. Next, the VR HMD was being fitted to the participants and the VR controllers being explained. Participants navigated through five different buildings: one for the tutorial condition and four for the actual trials. The four buildings were the same for every participant and in the same order. In the end room on each floor, participants had to sketch the shortest possible way from start to finish on a 2-D map. Then, they had to press a button when they were finished. After pressing the button, the map showed the actual position of the end room. Depending on whether it was the first or second floor of the building, participants were either instructed to step into the elevator or a voiceover said, "Thank you." Instruction for the task were provided in text and via pre-recorded audio in VR before the practice started, as well as when reaching and completing the final rooms in the actual trials. During the trials, participants were instructed to find and reach the target room as quickly as possible, as well as to sketch the map as quickly as possible. First, participants navigated the tutorial building as a practice trial where landmarks were present at junctions and the outdoor.

Then, they navigated through four other buildings where the presence of landmarks at junctions and outdoor landmarks varied. The order of the landmark conditions in buildings 1 to 4 was pseudorandomized and counterbalanced across participants and buildings. For instance, one participant might have no landmarks in the first building, outdoor landmarks in the second building, indoor landmarks in the third building, and both types of landmarks in the fourth building. The next participant would have the same conditions, but in a different order. This was done to counteract possible order effects. Other than the possible landmarks and cardinal directions (see Fig. 5a), there were no signs or other guidance for the participants on their way to the end room. Between Buildings 2 and 3, participants were required to pause and remove the HMD to prevent symptoms of cybersickness. After Building 4, they were debriefed. The VR part took about 60 to 70 min, including the break and questionnaires.

To evaluate the participants' orientation performance, we calculated the difference between their sketched maps and the shortest possible way, as well as the time it took to sketch. Therefore, we operationalized the opposite of accuracy as *"Inaccuracy of Orientation"*. The difference in accuracy of the map was calculated by adding up the Euclidean distances of each room of the shortest way possible to the corresponding part of the sketch map. To determine the corresponding part of the sketch map the following formula was used:

$$c_i = \frac{g}{r} * i \tag{1}$$

with c_i being the corresponding part for room number i on the shortest path, g being the number of marked chunks, and r being the number of rooms on the shortest path. For example, a participant marks seven chunks as the direct way on the upper floor, which in reality consists of only six rooms not counting the start room. The corresponding part for the first room, the one adjacent to the staring room is at $\frac{1}{6}$ between the center of the first and the second marked chunk. A visualization of this example can be seen in Fig. 6. In addition, the second outcome *"Time of Orientation"* for evaluating orientation performance was measured as the time required to sketch the map.

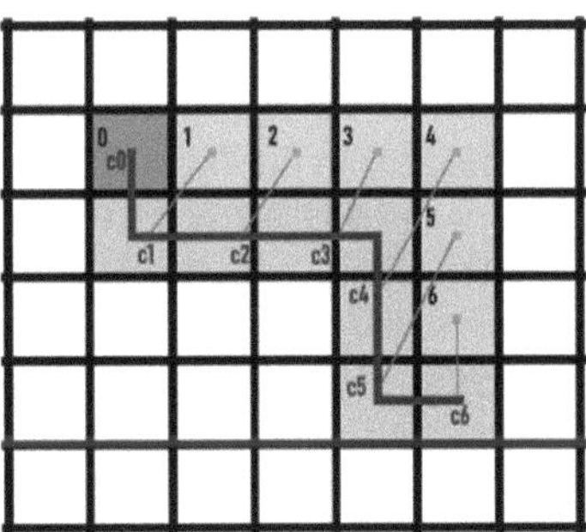

Fig. 6. Visualization of a sketch map. The start chunk of the floor is marked green, the shortest way is orange, and the user input blue. For each room 1–6 the corresponding point c_1-c_6 on the sketched path is calculated. To evaluate the sketched path, the Euclidean distances between the center of each correct chunk and the corresponding point are added up.

4.4 Statistical Analyses

We used IBM SPSS 28 for all statistical analyses. Sphericity was achieved for both dependent variables of performance (Mauchly tests all p's > .180). Overall, our design is a single-factor, within-subject design. The repeated factor, "landmark condition", was divided into four factor stages: no landmark, outdoor landmark, indoor landmark, and outdoor & indoor landmark. For this reason, we calculated repeated measures analyses of variance (ANOVA) with defined contrasts (hypotheses).

5 Results

Table 1 reports the descriptive statistics for the four experimental conditions and two dependent variables (performance measures of orientation). Figures 7 and 8 show the boxplots of each experimental condition. The data were normally distributed in all conditions (Kolmogorov-Smirnov p > .20).

Table 1. Descriptive Statistics for the four experimental conditions.

Measure	Without landmark		Landmark outdoor		Landmark indoor		Both: Indoor and outdoor	
	M	SD	M	SD	M	SD	M	SD
Time for orientation	42.23	21.30	32.68	22.57	43.82	28.79	43.00	29.88
Inaccuracy of orientation	14.44	11.63	8.54	9.39	10.51	10.30	8.38	9.05

Note: $N = 22$; M = Mean, SD = Standard Deviation.

In order to test our hypotheses, we examined the contrasts (a priori assumptions) using repeated measures Analysis of Variance (ANOVAs) for statistical analysis. To test hypothesis 1, we tested a contrast between the no landmark condition and all three landmark conditions together. We used the following contrast coding: 3–1 -1 -1. There was a statistically significant difference in the *"inaccuracy of orientation"* between the without landmark condition ($M = 14.44$, $SD = 11.63$) and the landmark conditions ($M = 9.14$, $SD = 5.64$), $F(1, 21) = 4.59$, $p = .044$ (see Fig. 9). This is a large effect, $\eta_p^2 = 0.179$ [49], which is statistically significant. The result shows that the absence of landmarks leads to greater *"inaccuracy of orientation"* than the presence of landmarks; therefore, hypothesis 1 can be maintained for the independent variable *"inaccuracy of orientation"*. In contrast, no differences were found for *"time for orientation"*.

Hypothesis 2 was also tested with a defined contrast in the context of repeated measures analysis of variance (ANOVA). The contrast was defined as follows: 0, -1, 1, 0. In line with the hypothesis, participants performed better in the outdoor landmark condition compared to the indoor landmark condition (see Fig. 7 and 8). However, these could not be statistically confirmed by the tested contrasts. There was no statistically significant difference in the *"time for orientation"* between the landmark outdoor condition ($M =$

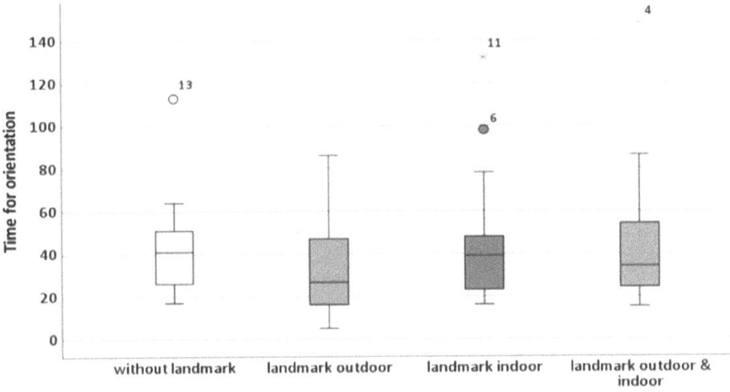

Fig. 7. Box plots for the four experimental within conditions and the dependent variable "Time for orientation".

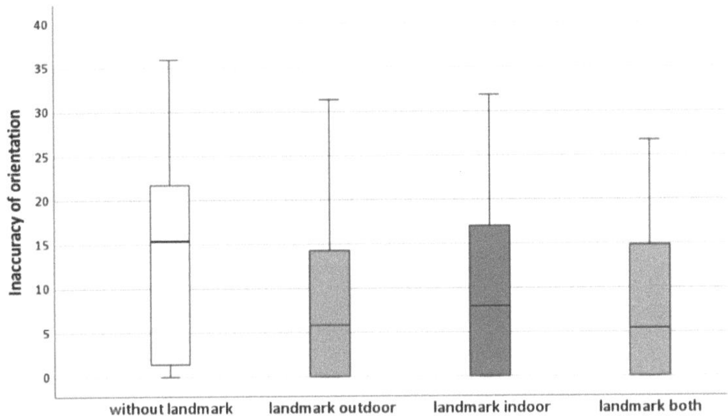

Fig. 8. Box plots for the four experimental within conditions and the dependent variable "Inaccuracy of orientation".

32.68, $SD = 22.57$) and the landmark indoor condition ($M = 43.82$, $SD = 28.79$), F $(1, 21) = 2.603$, $p = .122$, $\eta_p^2 = 0.11$. Similarly, there were no statistically significant differences in *"inaccuracy of orientation"*, between the outdoor landmark condition ($M = 8.54$, $SD = 9.39$) and the indoor landmark condition ($M = 10.51$, $SD = 10.30$), F $(1, 21) = 0.367$, $p = .551$, $\eta_p^2 = 0.017$. The necessary statistical test power was not achieved with the given sample size. For this reason, the hypothesis cannot be sustained.

Our third hypothesis assumed that, unlike the outdoor-only condition, participants in the outdoor & indoor landmark condition would need more *"time for orientation"*. In line with the hypothesis, participants showed no differences in *"inaccuracy of orientation"* concerning Euclidean Distance. There was no statistically significant difference in the *"inaccuracy of orientation"*, between the outdoor & indoor landmark condition ($M = 8.38$, $SD = 9.05$) and outdoor landmark condition ($M = 8.54$, $SD = 9.39$) F $(1, 21) = 0.005$, $p = .945$, $\eta_p^2 = .000$. However, we were unable to statistically confirm differences in

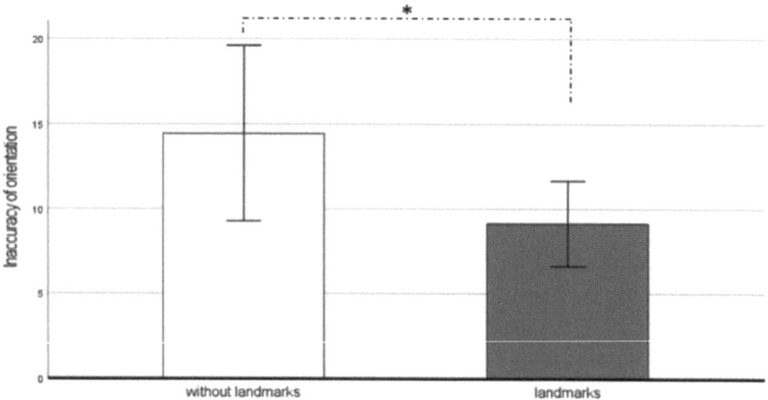

Fig. 9. Bar plots for contrasts for testing "without landmarks" condition vs. "landmarks" conditions. Error bars show 95% confidence intervals.

time required between the outdoor & indoor condition ($M = 43.00$, $SD = 29.88$) and the "outdoor" condition ($M = 32.68$, $SD = 22.57$), $F(1, 21) = 1.627$, $p = .216$, $\eta_p^2 = 0.072$. Although the non-existent difference in inaccuracy of orientation was consistent with the hypothesis (zero effect), the hypothesized effect on the time required for orientation assessments could not be statistically confirmed. The sample size was insufficient to statistically confirm the medium effect ($\eta_p^2 = 0.072$). For this reason, the hypothesis cannot be maintained at present.

6   Discussion and Conclusion

This study implemented a VR prototype of generated two-story buildings in which the participants needed to find the target room and sketch the shortest possible way to this room on a map at the end. According to hypothesis 1, it was predicted that participants perform better in the sketch mapping, if they have landmarks available for orientation. The results confirmed the assumption that landmarks are an important orientation factor. There was a significant difference in the accuracy of the maps ("*inaccuracy of orienta-tion*") between runs with and without landmarks in favor of having landmarks present (see Fig. 9). Participants did sketch more accurate maps in the conditions with land-marks than in the condition without landmarks. Building on the studies by Billinghurst and Weghorst [36] and Blades [58], this also suggests more accurate cognitive maps and environmental knowledge in these conditions. This result of our study is in line with previous studies on landmarks [31, 43]. Consequently, the present study contributes to the body of knowledge by demonstrating the efficacy of landmarks in navigating unfa-miliar and complex spaces. According to hypothesis 2, it was predicted that participants perform better in the orientation assessment, if outdoor landmarks are present instead of indoor landmarks. However, this hypothesis was rejected in our study because there was no statistically significant difference in the data to uphold it. Although the hypothesis could not be confirmed due to the statistical power, the data suggest that performance on the orientation assessment may be better when only outdoor landmarks are available.

This could be due to the aforementioned competition between the two types of landmarks, similarly to the competition between structural and object landmarks noted by Stankiewicz and Kalia [13].

According to hypothesis 3, the time required to sketch the map should be shorter when outdoor landmarks are present, compared to indoor and outdoor landmark conditions, while the map's accuracy should remain the same. Although this hypothesis could not be confirmed, the medium effect size and data from this sample seem to support this. However, Keskin et al. [38] did not find statistically significant differences between groups with different levels of expertise or between genders. Therefore, the fact that we found no time differences is consistent with the lack of difference in landmarks and sketch mapping in the study by Keskin et al. [38]. Since this was not part of our hypotheses, future research should further investigate our aspects of temporal performance in map sketching in an experimental setting to verify this.

Those trends towards the aforementioned effects, which were not statistically significant, could also be attributed to the sample size of this study. The sample size may also be a reason why this study did not replicate some of the differences between different types of landmarks of other studies, e.g., Stankiewicz and Kalia [13], Grzeschik et al. [30]. Although we found statistically significant differences according to our a-priori contrasts for hypothesis 1, as shown in Fig. 9, the differences between the different landmark conditions shown in Figs. 7 and 8, were not significant. An increased sample size, as well as a more streamlined experiment design, might lead to findings similar to those of Grzeschik et al. [30] and Stankiewicz and Kalia [13]. Another aspect that may have affected the lack of difference between different types of landmarks in our study is lighting and shadows. The lighting in our study was flat, without shadows. Natural lighting with shadows could have produced more noticeable, yet subtle, indirect directional cues due to size and shape differences in the large outdoor landmarks. This would have made a distinction between outdoor and indoor landmarks more apparent. Then, the outdoor landmark condition and the condition with the combination of both landmark types could have shown more tangible differences in performance compared to the other conditions. This would align with the findings of Grzeschik et al. [30] that more distinguishable landmarks lead to better navigation performance. A possible better performance in a lit environment with landmarks would also be in line with the findings of Marples et al. [59]. They found that participants completed the task faster in an environment with subtle guidance lighting and landmarks than in neutral lighting conditions, similar to those in our study. On the topic of more natural lighting and shadows, Cortes et al. [60] demonstrated that virtual embodiment increased and participants' spatial perception improved when shadows matched their body and movements. Although their study took place in a CAVE rather than an HMD VR environment, our study may have produced similar effects with natural lighting and accompanying body shadows rather than flat lighting and a lack of shadows. This could have led to more pronounced effects of the different types of landmarks and landmark conditions in our sample size. The lack of body shadows may also have affected embodiment in VR in our study. While literature on the combination of VR embodiment and spatial cognition is relatively scarce, the study by Knig et al. [61] suggest that better cognitive maps are formed in environment with a more embodied experience. However, our study did not set out to measure embodiment and its possible effect(s) on spatial cognition. Another aspect worth discussing is the general design aesthetics of our experiment. In this study we did not vary fidelity like Müller et al. [12] did in theirs. We decided to keep it at a medium level of fidelity with naturally looking textures being present, but not necessarily of high definition throughout all our used assets. Prior research [39, 62, 63], however, has shown that the fidelity and the experienced realism can influence different aspects of spatial cognition drastically in a positive way. This research does suggest that there might have either been a bigger effect or better overall performance if we would have used a level design that would be perceived more realistic. Using our approach to measuring cognitive mapping, future research could then also vary fidelity or use a high fidelity, high realism level design approach from the start to examine if our measurements might be influenced as well. Building on previous work [12, 43], indoor or locally arranged landmarks may serve as a reliable reference for navigation with few errors. Also building on the work of Ruddle et al. [43], outdoor or globally arranged landmarks do not introduce or reduce errors in navigation but seem to improve accuracy. Furthermore, having both types of landmarks simultaneously may make retrieving orientation information more difficult. It is possible that one type of landmark is more useful for retrieving orientation information, while the other type of landmark helps with navigation and orientation itself. One reason for this could be the higher cognitive load [50] caused by both types of landmarks, which results in a higher extraneous cognitive load. Another reason could be split attention [49] when both types of landmarks are present. Future research should address this issue, as our sample's data suggests this possibility, though it is not statistically significant.

Similar to Kim and Bock [21], this study attempted to make each building progressively more complex to avoid training effects. However, it did not necessarily add more decision points to the later buildings since the minimum number of rooms remained the same. However, this study added a loop to each upper floor to increase complexity, since the upper floor had fewer rooms than lower floor.

Nonetheless, the differences in performance that can be attributed to the presence of different types of landmarks are worth further investigating, particularly how these landmarks might facilitate traversing higher-than-ground floors. There may also be cases where different types of landmarks help in different parts of knowledge acquisition. VR makes it "easy" to implement these kinds of experiments regarding the environment as well as control confounding variables. VR also makes it rather easy to collect different kinds of behavioral data, such as task performance, reaction time, movement speed, and movement trajectories, all at once. Task performance measurements can be based on calculations similar to the method we described in the Methods section of this article. Doing so in the real world can be difficult or impossible, as Loveland and colleagues [7] mentioned. While this study used a prototype to randomly generate several buildings in advance that were then used across all participants, the next step could be to randomly generate buildings during the actual experiment. This could be based on factors established a priori or in an adaptive way, progressively making each trial more complex. Since Péruch et al. [8] and Dong et al. [10] argue that performance and spatial representation are very similar between the real world and virtual environments, the results of this study and further work in this area using VR have implications for the real world. Using salient landmarks, as in this study and in Ruddle et al.'s study [43], especially at possible decision points, can help navigate complex buildings. This could be helpful in designing emergency exit routes or guidance systems in a building with public businesses. Such a guidance system needs to be salient enough to help people find the room or exit they are looking for without needing prior training or understanding of the building's layout.

Disclosure of Interests. The author(s) declare no competing interests.

References

1. Qi, F., Lu, Z., Chen, Y.: Investigating the influences of healthcare facility features on wayfinding performance and associated stress using virtual reality. HERD. **15**(4), 131–151 (2022). https://doi.org/10.1177/19375867221108505
2. Chen, H., Zhi, J., Xiang, Z.-R., Zou, R., Ding, T.: Visualization analysis of emergency exit signs literature based on CiteSpace. Buildings. **13**(10), 2497 (2023). https://doi.org/10.3390/buildings13102497
3. Filippidis, L., Xie, H., Galea, E.R., Lawrence, P.J.: Exploring the potential effectiveness of dynamic and static emergency exit signage in complex spaces through simulation. Fire Saf. J. **125**, 103404 (2021). https://doi.org/10.1016/j.firesaf.2021.103404
4. Jamshidi, S., Ensafi, M., Pati, D.: Wayfinding in interior environments: an integrative review. Front. Psychol. **11**, 549628 (2020). https://doi.org/10.3389/fpsyg.2020.549628
5. Yesiltepe, D., Conroy Dalton, R., Ozbil Torun, A.: Landmarks in wayfinding: a review of the existing literature. Cogn. Process. **22**, 369–410 (2021). https://doi.org/10.1007/s10339-021-01012-x
6. Zuo, Y., Zhou, J.: Reducing younger and older adults' spatial disorientation during indoor-outdoor transitions: effects of route alignment and visual access on wayfinding. Behav. Brain Res. **465**, 114967 (2024). https://doi.org/10.1016/j.bbr.2024.114967
7. Loveland, T.R., et al.: Seasonal land-cover regions of the United States. Ann. Assoc. Am. Geogr. **85**, 339–355 (1995). https://doi.org/10.1111/j.1467-8306.1995.tb01798.x
8. Péruch, P., Belingard, L., Thinus-Blanc, C.: Transfer of spatial knowledge from virtual to real environments. In: Lecture Notes in Computer Science, pp. 253–264. Springer Berlin Heidelberg, Berlin, Heidelberg (2000). https://doi.org/10.1007/3-540-45460-8_19
9. Witmer, B.G., Bailey, J.H., Knerr, B.W., Parsons, K.C.: Virtual spaces and real world places: transfer of route knowledge. Int. J. Hum. Comput. Stud. **45**, 413–428 (1996). https://doi.org/10.1006/ijhc.1996.0060
10. Dong, W., et al.: Wayfinding behavior and spatial knowledge acquisition: are they the same in virtual reality and in real-world environments? Ann. Am. Assoc. Geogr. **112**, 226–246 (2021). https://doi.org/10.1080/24694452.2021.1894088
11. Jaksties, A., Sünderkamp, J.-H., Plümer, J.H., Müller, K.: Lost in 3d. In: Preim, B. (ed.) Proceedings of the Conference on Mensch Und Computer, pp. 133–136. ACM Digital Library, Association for Computing Machinery, New York, NY, United States (2020). https://doi.org/10.1145/3404983.3410013
12. Müller, J., et al.: The art of orientation - how not to be lost in 3d. In: Schneegass, S., Pfleging, B., Kern, D. (eds.) Mensch Und Computer, pp. 426–431. ACM, New York, NY, USA (2021). https://doi.org/10.1145/3473856.3473988
13. Stankiewicz, B.J., Kalia, A.A.: Acquistion of structural versus object landmark knowledge. J. Exp. Psychol. Hum. Percept. Perform. **33**, 378–390 (2007). https://doi.org/10.1037/0096-1523.33.2.378

14. Landau, B.: Spatial cognition. In: Encyclopedia of the Human Brain, vol. 395–418. Elsevier (2002). https://doi.org/10.1016/b0-12-227210-2/00326-5
15. Association, A. P. The Apa Dictionary of Psychology: Spatial Orientation (n.d.)
16. Siegel, A.W., White, S.H.: The development of spatial representations of large-scale environments. In: Advances in Child Development and Behavior, pp. 9–55. Elsevier (1975). https://doi.org/10.1016/s0065-2407(08)60007-5
17. Thorndyke, P.W., Goldin, S.E.: Spatial learning and reasoning skill. In: Spatial Orientation, pp. 195–217. Springer US, Boston, MA (1983). https://doi.org/10.1007/978-1-4615-9325-6_9
18. Darken, R., Peterson, B.: Spatial orientation, wayfinding, and representation. In: Handbook of Virtual Environments, 2nd edn. CRC Press (2014)
19. Montello, D.R.: A new framework for understanding the acquisition of spatial knowledge in large-scale environments. In: Spatial and Temporal Reasoning in Geographic Information, 143–154. Oxford University Press, New York, NY (1998)
20. Jansen-Osmann, P., Fuchs, P.: Wayfinding behavior and spatial knowledge of adults and children in a virtual environment. Exp. Psychol. **53**, 171–181 (2006). https://doi.org/10.1027/1618-3169.53.3.171
21. Kim, K., Bock, O.: Acquisition of landmark, route, and survey knowledge in a wayfinding task: in stages or in parallel? Psychol. Res. **85**, 2098–2106 (2020). https://doi.org/10.1007/s00426-020-01384-3
22. Taylor, H.A., Naylor, S.J., Chechile, N.A.: Goal-specific influences on the representation of spatial perspective. Mem. Cogn. **27**, 309–319 (1999). https://doi.org/10.3758/bf03211414
23. von Stülpnagel, R., Steffens, M.C.: Active route learning in virtual environments: disentangling movement control from intention, instruction specificity, and navigation control. Psychol. Res. **77**, 555–574 (2012). https://doi.org/10.1007/s00426-012-0451-y
24. Passini, R.: Wayfinding: a conceptual framework. Urban Ecol. **5**, 17–31 (1981). https://doi.org/10.1016/0304-4009(81)90018-8
25. Freundschuh, S.: Wayfinding and navigation behavior. In: Smelser, N.J., Baltes, P.B. (eds.) International Encyclopedia of the Social and Behavioral Sciences, pp. 16391–16394, Pergamon, Oxford (2001). https://doi.org/10.1016/B0-08-043076-7/02495-5
26. Allen, G.L.: Spatial abilities, cognitive maps, and wayfinding. Wayfinding Behav. Cogn. Mapp. Spat. Process. **4680**, 46–80 (1999)
27. Moura, D., Bartram, L.: Investigating players' responses to wayfinding cues in 3d video games. In: CHI '14 Extended Abstracts on Human Factors in Computing Systems, CHI EA, vol. 14, pp. 1513–1518. Association for Computing Machinery, New York, NY, USA (2014). https://doi.org/10.1145/2559206.2581328
28. Miller, J., Carlson, L.: Selecting landmarks in novel environments. Psychon. Bull. Rev. **18**, 184–191 (2010). https://doi.org/10.3758/s13423-010-0038-9
29. Keller, A.M., Taylor, H.A., Brunyé, T.T.: Uncertainty promotes information-seeking actions, but what information? Cogn. Res. Princ. Implic. **5**, 42 (2020). https://doi.org/10.1186/s41235-020-00245-2
30. Grzeschik, R., et al.: From repeating routes to planning novel routes: the impact of landmarks and ageing on route integration and cognitive mapping. Psychol. Res. **85**, 2164–2176 (2021). https://doi.org/10.1007/s00426-020-01401-5
31. Wang, L., Mou, W., Sun, X.: Development of landmark knowledge at decision points. Spat. Cogn. Comput. **14**, 1–17 (2013). https://doi.org/10.1080/13875868.2013.784768
32. Tolman, E.C.: Cognitive maps in rats and men. Psychol. Rev. **55**, 189–208 (1948). https://doi.org/10.1037/h0061626
33. Warren, W.H., Rothman, D.B., Schnapp, B.H., Ericson, J.D.: Wormholes in virtual space: from cognitive maps to cognitive graphs. Cognition. **166**, 152–163 (2017). https://doi.org/10.1016/j.cognition.2017.05.020

34. Weisberg, S.M., Newcombe, N.S.: Cognitive maps: some people make them, some people struggle. Curr. Dir. Psychol. Sci. **27**, 220–226 (2018). https://doi.org/10.1177/0963721417744521

35. Peer, M., Brunec, I.K., Newcombe, N.S., Epstein, R.A.: Structuring knowledge with cognitive maps and cognitive graphs. Trends Cogn. Sci. **25**, 37–54 (2021). https://doi.org/10.1016/j.tics.2020.10.004

36. Billinghurst, M., Weghorst, S.: The use of sketch maps to measure cognitive maps of virtual environments. In: Proceedings Virtual Reality Annual International Symposium, vol. 95, pp. 40–47. IEEE Comput. Soc. Press (1995). https://doi.org/10.1109/vrais.1995.512478

37. Rovine, M.J., Weisman, G.D.: Sketch-map variables as predictors of way-finding performance. J. Environ. Psychol. **9**, 217–232 (1989). https://doi.org/10.1016/s0272-4944(89)80036-2

38. Keskin, M., Ooms, K., Dogru, A.O., Maeyer, P.D.: Digital sketch maps and eye tracking statistics as instruments to obtain insights into spatial cognition. J. Eye Mov. Res. **11** (2018). https://doi.org/10.16910/jemr.11.3.4

39. Meijer, F., Geudeke, B.L., van den Broek, E.L.: Navigating through virtual environments: visual realism improves spatial cognition. Cyberpsychology Behav. Impact Internet, Multimedia Virtual Reality Behav. Soc. **12**, 517–521 (2009). https://doi.org/10.1089/cpb.2009.0053

40. Weisberg, S.M., Schinazi, V.R., Newcombe, N.S., Shipley, T.F., Epstein, R.A.: Variations in cognitive maps: understanding individual differences in navigation. J. Exp. Psychol. Learn. Mem. Cogn. **40**, 669–682 (2014). https://doi.org/10.1037/a0035261

41. Ruddle, R.A., Payne, S.J., Jones, D.M.: Navigating large-scale virtual environments: what differences occur between helmet-mounted and desk-top displays? Presence: Teleoperators Virtual Environ. **8**, 157–168 (1999). https://doi.org/10.1162/105474699566143

42. Jansen-Osmann, P.: Using desktop virtual environments to investigate the role of landmarks. Comput. Hum. Behav. **18**, 427–436 (2002). https://doi.org/10.1016/S0747-5632(01)00055-3

43. Ruddle, R.A., Volkova, E., Mohler, B., Bülthoff, H.H.: The effect of landmark and body-based sensory information on route knowledge. Mem. Cogn. **39**, 686–699 (2011). https://doi.org/10.3758/s13421-010-0054-z

44. Vilar, Elisângela, Rebelo, Francisco. Virtual Reality in Wayfinding Studies (2010). https://doi.org/10.1201/EBK1439834916-c80

45. Ewart, I. J., Johnson, H.: Virtual reality as a tool to investigate and predict occupant behaviour in the real world: the example of wayfinding, ITcon. Vol. 26, pp. 286–302 (2021). https://doi.org/10.36680/j.itcon.2021.016

46. Cen, D., Teichert, E., Hodgetts, C.J., et al.: Curiosity shapes spatial exploration and cognitive map formation in humans. Commun. Psychol. **2**, 129 (2024). https://doi.org/10.1038/s44271-024-00174-6

47. Zhang, X., Iwaki, S.: Self-location and reorientation of individuals without Reading maps: increased spatial memory during GPS navigation using AR City walls. In: Wei, J., Margetis, G. (eds.) Human-Centered Design, Operation and Evaluation of Mobile Communications. HCII 2024 Lecture Notes in Computer Science, vol. 14737. Springer Cham (2024). https://doi.org/10.1007/978-3-031-60458-4_17

48. Zuo, Y., Zhou, J.: Reducing younger and older adults' spatial disorientation during indoor-outdoor transitions: effects of route alignment and visual access on wayfinding. Behav. Brain Res. **465**, 114967 (2024). https://doi.org/10.1016/j.bbr.2024.114967

49. Tarmizi, R.A., Sweller, J.: Guidance during mathematical problem solving. J. Educ. Psychol. **80**, 424 (1988)

50. Sweller, J., Van Merrienboer, J.J., Paas, F.G.: Cognitive architecture and instructional design. Educ. Psychol. Rev. **251–296** (1998)

51. Shahid, A., Wilkinson, K., Marcu, S., Shapiro, C.M.: Karolinska sleepiness scale (KSS). In: Shahid, A., Wilkinson, K., Marcu, S., Shapiro, C.M. (eds.) STOP, THAT and One Hundred Other Sleep Scales, pp. 209–210. Springer, New York, New York, NY (2012). https://doi.org/10.1007/978-1-4419-9893-4_47

52. Kennedy, R.S., Lane, N.E., Berbaum, K.S., Lilienthal, M.G.: Simulator sickness questionnaire: an enhanced method for quantifying simulator sickness. The Int. J. Aviat. Psychol. **3**, 203–220 (1993). https://doi.org/10.1207/s15327108ijap0303_3

53. Hart, S.G., Staveland, L.E.: Development of NASA-TLX (task load index): results of empirical and theoretical research. In: Human Mental Workload, Vol. 52 of Advances in Psychology, pp. 139–183. Elsevier (1988). https://doi.org/10.1016/S0166-4115(08)62386-9

54. Pai, Y.S., Kunze, K.: Armswing: using arm swings for accessible and immersive navigation in ar/vr spaces. In: Proceedings of the 16th International Conference on Mobile and Ubiquitous Multimedia, MUM 2017, pp. 189–198. ACM (2017). https://doi.org/10.1145/3152832.3152864

55. Owl, E. N. ElectricNightOwl/ArmSwinger (2023). Original-date: 2016–07-06T04:58:29Z

56. American Psychological Association. Ethical principles of psychologists and code of conduct (2017). https://www.apa.org/ethics/ code/. Accessed 06 Oct 2021

57. World Medical Association: World medical association declaration of Helsinki. JAMA. **310**, 2191 (2013). https://doi.org/10.1001/jama.2013.281053

58. Blades, M.: The reliability of data collected from sketch maps. J. Environ. Psychol. **10**, 327–339 (1990). https://doi.org/10.1016/s0272-4944(05)80032-5

59. Marples, D., Gledhill, D., Carter, P.: The effect of lighting, landmarks and auditory cues on human performance in navigating a virtual maze. In: Symposium on Interactive 3D Graphics and Games, I3D '20, pp. 1–9. ACM (2020). https://doi.org/10.1145/3384382.3384527

60. Cortes, G., Argelaguet, F., Marchand, E., Lécuyer, A.: Virtual shadows for real humans in a cave: influence on virtual embodiment and 3d interaction. In: Proceedings of the 15th ACM Symposium on Applied Perception, SAP '18, pp. 1–8. ACM (2018). https://doi.org/10.1145/3225153.3225165

61. König, S.U., et al.: Embodied spatial knowledge acquisition in immersive virtual reality: comparison to map exploration. Front. Virtual Real. **2** (2021). https://doi.org/10.3389/frvir.2021.625548

62. Kimura, K., et al.: Orientation in virtual reality does not fully measure up to the real-world. Sci. Rep. **7**, 18109 (2017). https://doi.org/10.1038/s41598-017-18289-8

63. Pastel, S., Huber, M., Klinker, G., Weller, J.: Comparison of spatial orientation skill between real and virtual environments. Virtual Reality. **26**, 961–974 (2022). https://doi.org/10.1007/s10055-021-00539-w

Evaluating Multimodal AI Interfaces
for Cognitive Performance in Design Education

Claudia Nass Bauer[1]([✉]) [iD], Achim Ebert[2] [iD], and Thomas Lachmann[2] [iD]

[1] Mainz University of Applied Sciences, Mainz, Germany
claudia.nass-bauer@hs-mainz.de
[2] RPTU University Kaiserslautern-Landau, Kaiserslautern, Germany
{achim.ebert,lachmann}@rptu.de

Abstract. As AI systems increasingly permeate design education, the dominant interaction paradigm (text-based chat) risks constraining cognitive engagement in complex, iterative design tasks. This work explores whether and how multimodal interaction methods (e.g., visual, auditory, haptic) enhance cognitive performance compared to traditional chat-based AI interfaces within human-centered design (HCD) education. Based on an initial targeted literature analysis, multimodal interfaces have demonstrated benefits such as reduced cognitive load, increased user engagement, improved learning outcomes, and enhanced collaborative processes. The current work is situated within a broader doctoral research project investigating how generative AI reshapes cognitive design processes in novice designers. Building on an earlier case study that demonstrated that chat interfaces often lead to superficial understanding and linear thinking, this article urges a rethinking of AI-assisted design education as a multimodal, situational, and didactically coordinated experience. The current workshop provides a platform to exchange frameworks and strategies for engineering multimodal, cross-device AI experiences that better serve cognitive growth and design literacy in the generative age.

Keywords: Generative AI · Cognitive Performance · Design Education · Design Literacy · Interaction Modalities

1 Introduction

The rapid rise of generative artificial intelligence (AI) has opened new opportunities for supporting learning processes in design education [1]. However, most AI tools available to students, such as chat-based assistants, rely heavily on textual chat-based interaction [2]. While convenient and accessible, these tools may constrain the depth and nature of cognitive engagement in design tasks. As human-centered design (HCD) emphasizes exploratory, intuitive, and often multimodal thinking, the reliance on a single mode of interaction can inhibit critical cognitive processes. This position paper discusses how multimodal AI interaction methods, including visual, auditory, and haptic components, have the potential to better support cognitive performance in design education than traditional chat-based interfaces.

© The Author(s), under exclusive license to Springer Nature Switzerland AG 2026
C. Fayollas et al. (Eds.): EICS 2025, LNCS 16511, pp. 161–166, 2026.
https://doi.org/10.1007/978-3-032-26051-2_12

This argument is grounded in an ongoing doctoral research project focused on the cognitive imprint of AI in design literacy. The goal is to investigate how AI alters the mental models, workflows, and decision-making strategies of novice designers and how educational settings can integrate these tools without undermining fundamental cognitive development.

2 Background

Design education fosters not only technical skills but also cognitive capabilities like problem framing, synthesis, externalization, and reflection. Studies in cognitive design theory have shown that experienced designers engage in abductive reasoning, parallel idea generation, and iterative sketching. These practices are underpinned by intuition, reflection, and graphicacy, all essential for managing the ambiguity inherent in HCD [3, 4]. These activities benefit from a natural multimodal environment, where dialog, physical spaces, paper and pens play a role [3, 4]. Traditional AI tools, particularly chat-based systems, often support a linear mode of reasoning that limits opportunities for lateral exploration and reflective iteration [2].

In a prior case study conducted at the Mainz University of Applied Sciences [5], design students integrated generative AI into various stages of the HCD-process. While AI supported early-stage ideation and content generation, the study reported that chat-based tools produced shallow insights and required lengthy interaction sequences to generate value. The lack of spatial, visual, or interactive affordances limited the students' ability to iterate or reframe problems fluidly.

The cognitive implications of tool design in learning contexts are well grounded in instructional design research. According to Cognitive Load Theory (Sweller, 1994), working memory has a limited capacity, and instructional design must manage three types of cognitive load: intrinsic (task-related), extraneous (presentation-related), and germane (learning-related) [6]. Sweller (1994) emphasizes that the way information is presented can influence cognitive load, particularly by reducing extraneous load and thereby freeing up working memory for learning. This theoretical stance aligns closely with Mayer's Cognitive Theory of Multimedia Learning (2005), which argues that learners perform better when information is presented using a combination of words and visuals, as long as the content does not exceed the learner's processing capacity [7].

Therefore, the higher-level work seeks to explore the cognitive imprint that generative AI leaves on novice designers. Specifically, it investigates how AI tools reshape learners' mental models, decision-making strategies, and creative routines within the design process as well as identifies how such tools can be constructively integrated into design education to support foundational cognitive development and design literacy.

3 Targeted Literature Analysis

To open up the research space, an initial targeted literature analysis examined comparative studies on multimodal versus chat-based AI interfaces in design education, guided by the research question: **To what extent does multimodal interaction improve cognitive performance compared to traditional chat-based interaction methods when**

using AI in the human-centered design process within design education? The targeted literature analysis was conducted using the Semantic Scholar database, focusing on empirical studies published between 2020 and 2025 that explored the cognitive effects of multimodal AI interaction in design education settings. Semantic Scholar is an artificial intelligence-driven academic search platform, which indexes research from major scholarly sources including the ACM Digital Library, IEEE Xplore, Springer Nature, Elsevier, and arXiv, thereby enabling semantically enhanced retrieval and analysis of scientific literature [8].

The search yielded 50 initial results using keywords such as "multimodal AI," "design education," "cognitive performance," and "human-centered design." These articles were screened based on predefined inclusion criteria: studies had to involve design students or educators, include AI systems with at least one non-textual modality (e.g., voice, visual, or haptic), and report measurable outcomes related to cognitive engagement, mental workload, learning performance, or design thinking skills. Following this screening process, ten peer-reviewed empirical studies were selected for in-depth ana-lysis. From each study, data were extracted on research design, participant characteristics, interaction modalities, cognitive outcome measures, and comparative evaluation methods.

Collectively, these studies provide evidence that multimodal AI systems offer a range of cognitive advantages over traditional chat-based interfaces, reinforcing the need for further research into how interaction modality influences cognitive performance in design learning contexts. Key findings include:

- **Reduced Cognitive Load:** Kaufman et al. (2024) and Liao et al. (2023) used the NASA Task Load Index to demonstrate significant reductions in mental and physical workload when using multimodal AI systems compared to traditional interfaces [9, 10]. For example, in Liao's study, haptic feedback during smartwatch design tasks led to greater precision (accuracy of 0.88 vs. 0.86) and reduced frustration.
- **Enhanced Learning Outcomes:** Kumar (2021) found substantial gains in learning performance when students engaged with a multimodal chatbot (text, emojis,
- affective feedback), reporting a large effect size (Cohen's d = 0.978) in experimental groups [11].
- **Improved Engagement and Collaboration:** Rezwana and Maher (2022) showed that students interacting with a multimodal co-creative AI (text, voice, avatar) exhibited significantly higher user engagement and collaborative fluency compared to those using a button-based UI [12].

  Cross-study observations suggest:

- Cognitive benefits are particularly strong when the interaction modality aligns with the task type.
- Visual and tactile modalities support spatial reasoning and iterative design tasks.
- Auditory modalities aid verbal reflection and ideation.
- Multimodal systems foster more dynamic, intuitive engagement with content.

These results establish a compelling case for further empirical work investigating the specific mechanisms and didactical implications of multimodal AI interaction in human-centered design education.

4 Discussion Toward a Research Agenda

Recent developments in generative artificial intelligence (AI) are reshaping how students engage with design processes, often in ways that occur informally and without structured pedagogical oversight. Empirical evidence and classroom observations indicate that novice designers are already incorporating these tools into their academic and creative workflows, frequently outside of curricula or institutional frameworks [1, 5]. This unsupervised integration signals a significant pedagogical shift, highlighting the need to understand how such tools impact the cognitive development of learners. As underscored by existing literature and prior case study results, generative AI systems have the potential to both support and disrupt core design capabilities, including synthesis, iteration, and reflection. The central educational challenge, therefore, is not whether to include these technologies, but how to purposefully integrate them into learning environments in a way that strengthens fundamental cognitive competencies in human-centered design.

Addressing this challenge requires a critical understanding of how AI systems influence not only workflows but also students' cognitive processes. They influence how students approach problems, frame ideas, and make creative decisions. These tools have the potential to reshape workflows in subtle but significant ways, sometimes accelerating ideation, but also risking the erosion of critical, reflective, and exploratory thinking. Students must be equipped not only with operational proficiency, but with the critical literacy to interrogate AI outputs, assess their value, and integrate them responsibly into their design process. Understanding the cognitive imprint of AI, i.e. how it shapes mental models, reasoning strategies, and learning trajectories, is essential to cultivating reflective and resilient design practitioners.

The reviewed studies illustrate that when multimodal AI tools are integrated into education, they appear in various formats, from chatbots and avatars to simulators and wearable haptics. Across these diverse implementations, one theme emerged consistently: successful integration requires pedagogical alignment. Systems that merely layer AI assistance onto existing workflows risk fragmenting cognitive engagement. By contrast, studies such as Kaufman et al. (2024) and Kumar (2021) demonstrate that when multimodal support is embedded contextually, whether through verbal cues, emotional feedback, or tactile interfaces, cognitive support is more effective and less intrusive [9, 11].

In designing AI-supported learning environments, special attention must be paid to the alignment between interaction modality and cognitive task. The literature suggests that different modalities afford distinct types of cognitive engagement, e.g. auditory modalities support brainstorming and verbal ideation, while visual scaffolds enhance synthesis, spatial reasoning, and systems thinking. Mapping these modality-task relationships more systematically could inform not only the development of more effective AI tools but also support curriculum design and instructional planning in design education.

These empirical insights also reflect longstanding principles in instructional design theory, particularly the theories of Sweller (CLT) and Mayer. Sweller (1994) argues that cognitive overload can be mitigated by distributing mental effort across distinct presentation modes, which frees up working memory for learning-relevant processes [6]. In this light, the modality-task alignment observed in studies such as Liao et al.

(2023) and Kaufman et al. (2024) can be interpreted as a way to reduce extraneous load while improving cognitive efficiency [9, 10]. Similarly, Mayer's framework suggests that learners benefit from multimedia input when information is presented through coordinated verbal and visual channels, as long as the format supports rather than overloads processing capacity [7]. These perspectives reinforce the idea that multimodal AI tools, if thoughtfully integrated into learning environments, can support cognitive performance rather than compete for limited mental resources. This may be especially relevant for novice designers, whose cognitive strategies and design heuristics are still developing and thus particularly sensitive to how instructional tools are structured.

To advance research on multimodal AI in design education, robust and multifaceted evaluation strategies are required. Future research should combine established cognitive assessment tools with research methodologies that capture the complex, iterative nature of human-centered design learning. First, validated instruments such as the NASA Task Load Index (TLX), used in studies by Kaufman et al. (2024) and Liao et al. (2023), are indicated for quantifying mental workload and user stress across different modalities [9, 10]. Second, performance-based measures, such as task accuracy (as used by [11]), output quality, and creativity rubrics [9, 10], provide direct indicators of cognitive outcomes like synthesis and problem-solving. Third, to capture more nuanced learning effects over time, researchers should implement longitudinal study designs, a gap identified across most reviewed studies, to observe how repeated exposure to multimodal AI influences learners' mental models and design strategies. Finally, mixed-methods evaluations that triangulate quantitative workload and performance data with qualitative insights (e.g., user reflections, interviews, screen recor-dings) would enrich understanding of how modality affects not just task performance, but also metacognition and engagement.

Taken together, these approaches offer a comprehensive yet flexible framework for studying the pedagogical impact of multimodal AI systems in HCD education.

5 Conclusion

In the era of generative AI, design education must evolve to reflect the cognitive and technological demands of practice. Multimodal AI interaction offers a promising avenue for enriching cognitive engagement, supporting intuitive exploration, and nurturing design literacy. This initial targeted literature analysis represents a first step in the broader doctoral project, providing a synthesized understanding of current findings, surfacing methodological insights, and showing some design parameters for future research tools.

By clarifying how multimodal systems differ cognitively from chat-based interfaces and identifying which modalities support which design tasks, the review equips the author with ideas for empirical planning. Ultimately, the findings from this early study will guide the construction of a research design that is not only methodologically robust but also pedagogically meaningful, ensuring the doctoral project's long-term contribution to both theory and practice.

Acknowledgments. The author thanks the students of the AI & Human-Centered Design course at Mainz University of Applied Sciences for their participation and insights, which inform this ongoing research.

166   C. Nass Bauer et al.

Disclosure of Interests. The author has no competing interests to declare that are relevant to the content of this article.

References

1. Chellappa, V., Luximon, Y.: Understanding the perception of design students towards ChatGPT. Comput. Educ.: Artif. Intell. **7**, 100281 (2024)
2. Chauhan, D., Singh, C., Rawat, R., Dhawan, M.: Evaluating the performance of conversational AI tools: a comparative analysis. In: Conversational Artificial Intelligence, pp. 385–409 (2024)
3. Cross, N.: Design Thinking: Understanding How Designers Think and Work. Bloomsbury Academic, London (2011)
4. Lawson, B.: How Designers Think: The Design Process Demystified. Routledge, London (2006)
5. Nass Bauer, C.: Exploring the integration of generative AI in user-centered design: a case study with design students. In: Mensch und Computer 2024 - Workshop Proceedings, Gesellschaft für Informatik e.V., Karlsruhe (2024). https://doi.org/10.18420/muc2024-mci-ws09-185
6. Sweller, J.: Cognitive load theory, learning difficulty, and instructional design. Learn. Instr. **4**(4), 295–312 (1994). https://doi.org/10.1016/0959-4752(94)90003-5
7. Mayer, R.E.: Cognitive theory of multimedia learning. In: Mayer, R.E. (ed.) The Cambridge Handbook of Multimedia Learning, pp. 31–48. Cambridge University Press, Cambridge (2005). https://doi.org/10.1017/CBO9781139547369.005
8. Lo, K., Wang, L.L., Neumann, M., Kinney, R., Weld, D.S.: S2ORC: the semantic scholar open research corpus. arXiv preprint arXiv:1911.02782 (2019)
9. Kaufman, R., Costa, J., Kimani, E.: Effects of multimodal explanations for autonomous driving on driving performance, cognitive load, expertise, confidence, and trust. Sci. Rep. **14**, 11294 (2024). https://doi.org/10.1038/s41598-024-62052-9
10. Liao, Y.-C., et al.: Interaction design with multi-objective Bayesian optimization. IEEE Pervasive Comput. **22**(1), 30–39 (2023). https://doi.org/10.1109/MPRV.2022.3230597
11. Kumar, J.A.: Educational chatbots for project-based learning: investigating learning outcomes for a team-based design course. Int. J. Educ. Technol. High. Educ. **18**, 38 (2021). https://doi.org/10.1186/s41239-021-00302-w
12. Rezwana, J., Maher, M.L.: Understanding user perceptions, collaborative experience, and user engagement in different human-AI interaction designs for co-creative systems. In: Proceedings of the Creativity & Cognition Conference. ACM (2022). https://doi.org/10.1145/3527927.3532789
13. Chen, T.-J., Krishnamurthy, V.R.: Investigating a mixed-initiative workflow for digital mind-mapping. J. Mech. Des. **142**(9), 091403 (2020). https://doi.org/10.1115/1.4046808

XAI Beyond Reality: Identifying Key Research Gaps and Future Directions

Kai Jonas Klingshirn(✉) ⓘ, Christoph Garth ⓘ, and Achim Ebert ⓘ

RPTU Kaiserslautern-Landau, Kaiserslautern, Germany
{k.klingshirn,garth,achim.ebert}@rptu.de

Abstract. Artificial Intelligence (AI) is increasingly being integrated into a wide array of Extended Reality (XR) applications, including sophisticated navigation systems, immersive training simulations for educational use-cases and data analysis in medicine, e.g. for MRI scans. Consequently, ensuring the transparency and interpretability of these AI-driven applications has become a major challenge. This paper examines the growing importance of Explainable AI (XAI) in Extended Reality environments and identifies key challenges in developing effective explanation systems. We analyze how these AI-powered XR applications particularly benefit from transparent explanations that build trust, enhance user understanding and improve overall adoption. After summarizing the general challenges in the field of XAI, we investigate how these challenges manifest in the specific context of XR. By synthesizing current research and identifying critical open questions, this work aims to guide future XAI development towards more transparent, trustworthy systems that prioritize human needs across XR applications and beyond.

Keywords: Challenges · Explainable AI · Extended Reality · Future Directions · Research Gaps · XAI · XR

1 Introduction

Extended Reality (XR), consisting of Augmented Reality (AR), Virtual Reality (VR) and Mixed Reality (MR), is rapidly advancing and increasingly being integrated into a wide range of domains, including safety-critical applications such as driver assistance, intelligent healthcare, educational training and medical diagnostics. Central to this development is the integration of Artificial Intelligence (AI), which enables XR systems to process vast amounts of data, adapt dynamically to user behavior and deliver context-aware, intelligent responses in real time. For example, AI algorithms can enhance driver assistance systems by processing real-time sensor data to support lane keeping, enable autonomous driving, or trigger emergency braking in the event of an imminent collision [25,39]. Similarly, fitness applications employ AI to monitor workout and health metrics, thereby providing personalized recommendations [21]. Particularly significant is the use of AI in the medical domain, where algorithms are applied to analyze for example CT scans of lungs and assist in disease diagnosis, ultimately

C. Fayollas et al. (Eds.): EICS 2025, LNCS 16511, pp. 167–184, 2026.
https://doi.org/10.1007/978-3-032-26051-2_13

improving patient care and treatment outcomes [10]. Such systems are commonly referred to as Clinical Decision Support Systems (CDSSs) [15]. As AI models grow in size and complexity, e.g. modern deep neural networks can contain billions of parameters, their internal workings become increasingly intransparent. Traditional methods, such as manually inspecting model behavior or analyzing input-output data, are no longer sufficient to achieve meaningful interpretability. Consequently, these models are frequently described as *black boxes*, as the reasoning behind their decisions remains largely inaccessible not only to end users but often even to their developers.

It is widely recognized that AI can significantly improve and simplify processes across a broad range of applications. As AI becomes more deeply integrated into XR environments, the complexity and autonomy of these systems continue to grow. While this integration unlocks unprecedented opportunities, it also raises one critical question: Can we trust the decision-making of an AI agent? For instance, understanding why an AI assistant issues a particular medical diagnosis is essential for doctors, who must decide whether to accept the recommendation and, if so, implement the corresponding treatment for the patient [10]. Addressing this challenge requires making AI decision-making processes more transparent and understandable. This is the central aim of the research field of eXplainable AI (XAI), which seeks to comprehensively visualize AI-driven decisions by approximating and elucidating their underlying mechanisms. Improved interpretability not only enhances trust in AI-supported systems but also increases their usability, thereby fostering much higher adoption.

The need for explainability extends across all AI-powered systems, regardless of application context, underscoring its role as a fundamental research priority. In the context of XR, this requirement is even more pronounced. XR systems often operate in dynamic, multi-sensory environments where users interact simultaneously with digital and physical elements in real time. The immersive and interactive nature of XR amplifies the impact of AI-driven decisions, making transparent and actionable explanations indispensable for effective human-computer collaboration [18,24].

Explanations can either focus on a specific prediction of an individual instance, or aim to understand the overall behavior and feature importance of the model across the entire dataset. Model-agnostic methods, such as *LIME (Local Interpretable Model-agnostic Explanations)* [38] and *SHAP (SHapley Additive exPlanations)* [23], analyze the decision-making based on a high number of input-output relationships, treating the AI system as a black box. In contrast, model-specific methods are tailored to the architecture of particular models, like feature importance derived directly from tree-based models. Furthermore, explanations can be ante-hoc, where the model is designed to be interpretable from the outset (e.g., decision trees), or post-hoc, where explanation techniques are applied after model training. The choice of XAI method depends on the application, complexity of the model, and level and scope of explanation [11].

According to social scientist Markus Langer and his colleagues [27], effective explanations in XAI must be deeply aligned with users' specific goals, prior knowledge and the contexts of use. The authors argue that explainability is not a one-size-fits-all property. Rather, explanations must be adaptive, reflecting the varying needs of different users depending on their roles, expertise levels and decision-making requirements. A key requirement is that explanations should be actionable, meaning they should not only clarify the reasoning behind outputs, but also empower users to make informed choices or interventions based on the information provided [35]. Furthermore, explanations must be evaluated against the actual outcomes they produce for users: whether they enhance understanding, facilitate trust or enable better decisions. To combine, we want explanations to be user-centered, goal-oriented and practically impactful [27].

Despite the increasing recognition of XAI as a crucial research field, it still faces substantial challenges, particularly in the context of XR. These challenges include the demand for real-time explanations, the integration of multimodal data and the need to balance transparency with user experience [18, 24].

By synthesizing insights from recent literature and identifying key research gaps, this work aims to guide future development of XAI towards systems that are more transparent, trustworthy and human-centered. Our analysis highlights the importance of designing explanation methods that not only clarify AI behavior but also foster user understanding, support informed decision-making and ultimately enable the responsible adoption of AI-powered XR technologies across diverse application domains.

1.1 Methodology

We provide a systematic review of XAI research challenges, emphasizing those within XR settings. Our method consisted of identification, screening and synthesis stages, targeting high-impact conferences such as ACM CHI and only included peer-reviewed papers published within the past five years. We mainly focused on IEEE and ACM digital libraries and searched for "explainable artificial intelligence", "explainable AI" and "XAI" resulting in a total of 1,362 publications. This initial result was filtered by adding the keywords "challenges", "gaps" and "open problems", followed by screening abstracts for relevance and excluding duplicates, irrelevant works and short papers lacking substantive contributions. Furthermore, we focused on papers summarizing or exploring current XAI challenges, which reduced the number of considered papers to 243 publications. For the XR part, we used the keywords "Extended Reality", "Augmented Reality", "Virtual Reality", "Mixed Reality" and their corresponding abbreviations, which resulted in 24 XAI-related papers in this subdomain. A final selection of over 200 publications was analyzed to synthesize key insights, resulting in a structured overview of the main XAI challenges both in general and within XR environments.

2 Key Obstacles in Achieving Explainable AI

Explainable AI faces a diverse set of challenges that span across multiple disciplines, like law and regulations, human-centered design, model engineering and ethical dimensions. These obstacles hinder the development and deployment of transparent, interpretable and trustworthy AI systems. In this section, we provide a structured overview of the most critical issues in recent literature, including methodological limitations, user adaptation issues, explanation evaluation difficulties and issues related to user trust and misuse. By systematically examining these key obstacles, we lay the foundation for understanding the unique requirements and open questions that arise when applying Explainable AI in complex, safety-critical environments such as education and medicine.

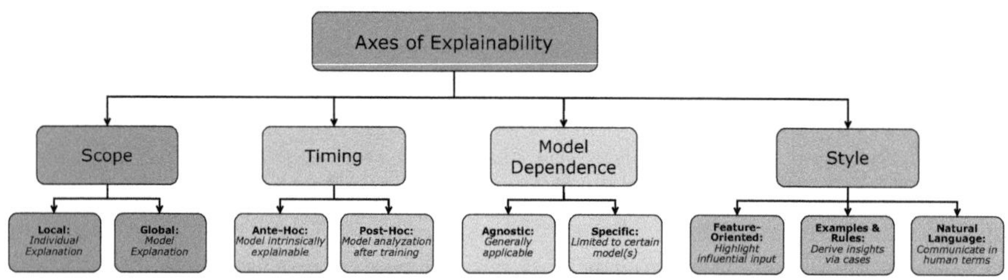

Fig. 1. The Four Axes of Explainability: *A framework for classifying and understanding different types of Explainable AI methods based on four key dimensions: scope, timing, model dependence and style. The Figure is based on categories introduced by Langer et al.* [27].

As visible in Fig. 1, explaining AI decisions can be approached through various methodologies, which can be categorized on four axes: (1) *scope*, (2) *time of explanation extraction*, (3) *dependence on the underlying model* and (4) *style*. The first dimension is *scope*, which refers to whether the explanation focuses on explaining a specific decision (local) or tries to approximate the overall model behavior (global). The *timing of explanation* extraction, can be either ante-hoc, where the model is designed to be interpretable from the outset, or post-hoc, where explanation techniques are applied after model training. A more intuitive description of ante-hoc explainability would be that these kind of explanations are built into the model from the beginning, making the model intrinsically interpretable. Post-hoc explanation methods analyze the relationship between inputs and outputs to infer insights about the internal workings of an AI model. These insights are then used to construct explanations for the model's decisions, without requiring access to the model's internal structure or parameters. Overall, some approaches are *dependent on the underlying model*, while others are model-agnostic and can therefore be generically applied to a wide variety of AI algorithms. The last important characterization is the *style* of explanations: they can be feature-oriented data visualizations highlighting the most influential factors, provide similar example cases, construct decision rules or provide natural language explaining the inner workings of an AI algorithm. Although there is no standardized classification, many papers on XAI build on this taxonomy. This highlights the first research gap in XAI: the lack of a consistent terminology that facilitates clear and effective discourse within the field.

In addition to the *axes of explainability*, five general categories of XAI methodologies can be distinguished, each representing a distinct strategy for elucidating the internal behavior of black-box models to enhance transparency.

The first group of approaches is called *Visual and Surrogate Models*. They aim to make complex models interpretable by either visualizing their decision logic or approximating them with simpler models. Visualization tools such as partial dependence plots and parallel coordinate plots offer insights into a model's behavior. Surrogate models, like decision trees or linear regression, are trained to mimic the full model's output, providing a simpler interpretable proxy [16].

Another collection of approaches can be summarized under the term *Feature Attribution Methods and Local Explanations*. These techniques assign importance scores to individual input features, helping to identify which aspects of the input were most influential in the model's decision-making process. A good example of this category is *LIME* (Local Interpretable Model-agnostic Explanations) [38], which locally approximates the model around a specific instance.

The next category of techniques are *Contrastive and Counterfactual Explanations*. These methods focus on understanding model decisions by exploring alternative scenarios. Contrastive explanations clarify why a specific outcome was chosen over another, while counterfactual explanations identify the minimal changes needed to alter the decision. This provides the user with a great understanding of a specific decision and its boundaries.

Rule-based and Example-based Explanations are another group of techniques that provide interpretability by using explicit if-then rules to define decision boundaries or by citing similar past cases to justify predictions [34, 47].

The last group of XAI methods are *Natural Language Explanations*. These systems have been evolving recently and express how models make decisions in natural language, leveraging templates, retrieval mechanisms or question-answering frameworks. Recently, Large Language Models (LLMs) have been investigated as tools for generating context-aware, interactive conversational explanations [3].

Having provided a general overview of the various XAI methods, we now examine in greater detail several of the underlying algorithms that implement these approaches. The *LIME* algorithm tries to locally explain a specific decision by approximating the model around that specific instance [38]. *LIME* therefore generates a set of perturbed copies of the original data point and feeds them into the black box model to predict each of them. It then trains a lightweight, interpretable surrogate, usually a linear regression. The coefficients of that model are similar to the original model within that local scope and reveal which features most strongly influence that particular prediction. *SHAP* on the other hand is an approach based on game theory and tries to show each feature's contribution to a specific prediction [23]. *SHAP* calculates a so called *SHapley value* for every feature, which describes the amount of influence of the feature on the prediction. To calculate that value, the algorithm evaluates how the model output changes when that feature is added to or removed from all possible feature subsets. By averaging over many background samples, *SHAP* provides a consistent, locally accurate attribution for each feature in the individual prediction. A further well-established XAI algorithm is *Anchors*, which tries to identify minimal conditions (rules) that guarantee a specific prediction with high probability [37]. Anchors starts with an empty set of feature-value conditions and iteratively adds the condition that increases the precision of the rule the most, using sampling to estimate how often the rule still produces the same prediction on perturbed instances. Once a rule reaches a user-specified precision, the algorithm prunes it to remove any unnecessary conditions while keeping the precision. In the end, the algorithm provides a set of high-confidence rules that explain the prediction. Finally, we consider *Grad-CAM*, which stands out as a key XAI technique for image classification tasks. It illuminates the "why" behind a Convolutional Neural Network's prediction by generating a visual heatmap. This heatmap is derived from the gradients of the target class back-propagated to the final convolutional layer, weighting each feature map by its importance and then summing them to highlight the most influential image regions for the decision [41].

In addition to these core algorithms, numerous extensions have been developed, such as *DLIME* [49], *Deep SHAP* [23], *Kernel SHAP* [23] and *Grad-CAM++* [7]. While these methods provide technically sound explanations, they are hard to interpret for laypeople, because of a lack of either machine learning or domain knowledge [42]. Even though the XAI toolset is quite effective and provides meaningful explanations, these algorithms and their associated visualizations cannot fully satisfy all requirements for explanations. One of the most important aspects is the need for human-centered explanations, which are easy to understand for all users, especially non-experts. In the following, we examine current research efforts aimed at improving XAI explanations to enhance their clarity and effectiveness. Figure 2 provides an overview of the primary research challenges in general XAI, highlighting XAI in XR as a subset within these challenges.

Social science research by Miller [29] contributes to addressing these challenges by emphasizing that explanations provided by AI agents should not be merely factual descriptions, but should also be contrastive, selective and social in nature. Humans tend to ask contrastive questions, such as "Why P rather than Q?" and prefer concise explanations that highlight key causes over exhaustive lists of contributing factors. Furthermore, people favor explanations grounded in concrete causes rather than abstract probabilities or statistical regularities. Even if a prediction is highly probable, a simple causal explanation is often perceived as more satisfying. Explanations are also inherently selective; they do not include all possible causes but instead focus on those deemed most relevant or salient within a given context. Additionally, explanations function as conversational artifacts, shaped by the explainer's understanding of the explainee's beliefs, intentions and goals. This emphasizes the importance of an explanatory self-model within AI systems, enabling them to adapt and tailor their explanations to a user's needs and level of understanding. Because of these human tendencies, Explainable AI must do more than just show a model's internals. It has to be able to identify a user's contrastive questions, select a small, context-relevant set of causes and present them in a conversational, causal form that builds trust.

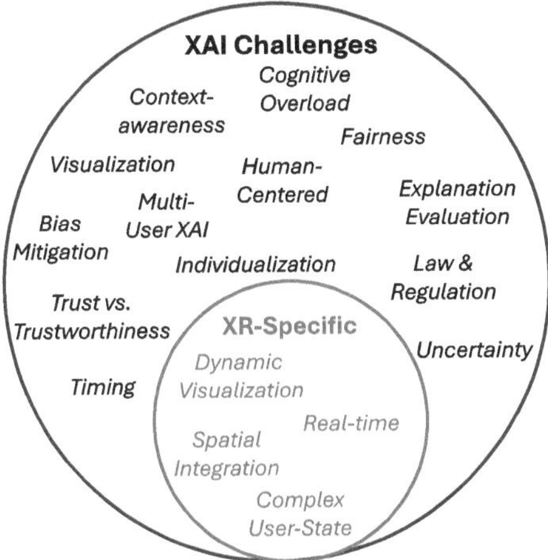

Fig. 2. General and XR-specific research challenges of XAI: *A comprehensive overview of the most relevant research challenges concerning explainable AI, encompassing both general difficulties and those unique to extended reality environments.*

Based on these findings, a central question emerges: what constitutes a good explanation and how can it be formally defined? If such a static definition existed, it could enable the automatic verification of an explanation's usefulness, comprehensibility and correctness ensuring that users are consistently presented with the most suitable explanation. Shajalal et al. [42] have explored this topic in depth, arguing that the form and structure of explanations can be analyzed across multiple levels: *syntax, semantics* and *pragmatics*. The syntax level refers to the presentation and visual encoding of explanations, emphasizing user-friendly choices in colors, fonts, layouts and chart design to minimize cognitive load. The semantic level describes how explanations are interpreted and which mental models are generated. Explanations must be mathematically precise and assist users in developing accurate mental models. The pragmatic level addresses the context-dependent practical significance of explanations as they apply to the users' daily lives [42]. Moreover, providing explanations that are too complex or detailed lead to a higher processing time or even to users skipping the explanation, lowering the system's usability or making it impractical to use [27]. Crafting a well-structured and therefore useful explanation requires careful consideration of all the aspects discussed above.

Beyond the visual representation, timing presents another crucial aspect [46]. Determining when to provide explanations (before, during or after AI decisions) and how they might evolve over time as both the context and user understanding develop throughout the interaction is definitely an aspect that should be further addressed [8]. In addition, prior research has shown that "[...] explanations should not always be presented to users, because they can introduce unnecessary cognitive load and become overwhelming [...]" [46]. This point of balance varies significantly based on user expertise, task criticality and time constraints. Real-time scenarios demand concise but powerful explanations, while non-time-sensitive scenarios permit multiple explanations providing broader contextual understanding. For example, in smart home environments, users often interact through their smartphones or tablets with limited screen space, necessitating the design of explanations for small-screen visualizations and good timing [42].

To ensure effective human-computer interaction, explanations must be easily understandable for all kinds of users, e.g. different expertise levels, cultural backgrounds, contexts and goals. Thus, tailoring explanations to individual needs and backgrounds is a critical requirement [1,18,20,29,31,42]. Current research approaches typically only offer, as we call it, "static personalization" in a group level, meaning the explanation is tailored to a specific user group, like domain experts or machine learning experts [30]. "The majority of XAI systems are non-adaptive. They provide static explanation types without dynamically adjusting to user profiles or context" [30]. Due to that, explanations are still not sufficient to achieve the required personalization level for universal effectiveness. A truly effective system has to be able to create human-centered explanations that incorporate not only individuals' knowledge but also their perception and biases [1,29,30]. It is our position that, beyond tailoring explanations to individual users, explanations should dynamically evolve as a user's knowledge and experience grow. Static user profiles or fixed knowledge groups are insufficient. Effective XAI systems should continuously adapt explanations to a user's evolving understanding and their needs over time, while also tailoring them to the specific context and goal of the interaction. According to Langer et al. [27], the usefulness of an explanation depends heavily on the context in which it is received. This context arises from the interaction between stakeholders, the system, the task and the environment. Together with factors such as time pressure or workload, these elements determine the level of understanding required.

Moreover, uncertainty visualization is also an important aspect of research, as it enhances transparency by conveying the confidence level of predictions and therefore helps users interpret the explanations provided. In addition to that, Prabhudesai et al. [36] found that visualizing uncertainty makes the users think analytically about the presented decisions and reduces the over-reliance on decision-support-systems. Traditional XAI methods often produce deterministic explanations, which may contribute to user overtrust. Incorporating uncertainty can help users better assess the reliability of predictions, which is an essential consideration in high-stakes domains. However, the complexity of the visualization should not overwhelm users and remain clear and actionable.

Providing accurate explanations is paramount, as incorrect explanations can significantly erode user trust and undermine the credibility of the entire system [22]. When considering trust in a system, it is important to distinguish between trust and trustworthiness. As Kästner et al. [22] note, "Trust is an attitude a stakeholder holds towards a system. Trustworthiness, by contrast, is a property of a system: intuitively, a system is trustworthy for a stakeholder when it is warranted for the stakeholder to put trust in the system." Building for trustworthiness, ensuring systems are fair, robust and transparent, should take priority, because trust itself depends on unpredictable factors such as a user's mindset, general attitude towards the system, prior experience with similar systems and social environments' attitude toward such systems. "If designers neglect trustworthiness and build an untrustworthy system, we will likely have either an untrustworthy system that most stakeholders will not trust in the long run or an untrustworthy system that is trusted mistakenly, with potentially devastating consequences" [22]. Even accurate explanations can erode trust if they reveal undesirable system behavior or are too complex to understand. The authors identify three reasons why explanations might fail to foster trust: (1) If trust is already maximal, (2) If the explanation reveals system problems, or (3) If the explanation cannot be properly comprehended. Unfortunately, such understandings of trust may be manipulated for malicious purposes, with some systems leveraging XAI to create a false impression of transparency. Chromik et al. [9] warn about "dark patterns", user interface strategies that simulate explainability without offering real insight, deceiving users into misplaced trust. Similarly, Brunotte et al. [4] caution against "XAI greenwashing", where superficial explanations are used to obscure ethical or technical deficiencies. These practices highlight the need for standards and regulations to ensure explanation integrity.

In addition to that, AI systems enter heavily regulated fields and high-stakes decision contexts, with the most prominent examples being medicine, finance and justice, which also contributes to the need for regulations and standards. In these environments, explanations must comply with specific legal, ethical and regulatory requirements. One key provision is the "right to explanation" within the General Data Protection Regulation (GDPR) [12], enacted by the European Union in 2016. This right grants individuals meaningful information about the logic, significance and consequences of automated decisions affecting them, ensuring transparency and accountability in automated decision-making processes. The GDPR itself is a comprehensive regulation governing data protection and privacy across the European Union and the European Economic Area. Established by the European Parliament and the Council of the European Union, it aims to protect individuals' personal data, grant them control over its use, harmonize data privacy laws across Europe and foster a safer digital environment. More recently, the EU AI Act [13], introduced in 2024, further strengthens these requirements by setting detailed rules for the deployment and governance of AI systems within the EU. These regulations emphasize the need for transparency and explanations that meet strict legal, ethical and compliance standards.

Beyond legal and regulatory frameworks, ethics in XAI focus on ensuring that AI systems operate with transparency, fairness and accountability in their decision-making processes. Bias, discrimination and lack of accountability can have severe consequences, e.g. in the medical field [44]. By enhancing model interpretability, XAI empowers stakeholders to understand and challenge AI-driven decisions, thereby safeguarding individual rights and ensuring compliance with legal frameworks such as the GDPR [12]. AI systems often inherit biases embedded in historical training data, which can reflect and perpetuate existing societal inequalities. These biases may cause sensitive attributes, such as race or gender, to influence predictions, either directly or indirectly. Zhou et al. [50] emphasize that explanations should help identify the impact of such sensitive features to promote fairness and accountability in decision-making.

Beyond regulatory considerations, researchers must develop standardized metrics and evaluation frameworks to assess the effectiveness of XAI methods prior to their deployment in real-world scenarios. Additionally, it is essential to validate whether these XAI approaches maintain their efficacy across diverse application domains. The need for such standards is paramount for several reasons: Firstly, standardized metrics enable meaningful comparisons between different XAI methods, allowing researchers and practitioners to objectively assess which techniques perform best under various conditions. Secondly, standards foster trust and accountability by providing a consistent way to evaluate the reliability and fidelity of explanations. Without them, claims of explainability can be subjective and difficult to verify. Moreover, standards can facilitate regulatory compliance by offering a benchmark against which the adequacy of explanations in regulated and high-stakes applications can be assessed. Finally, standardized evaluation frameworks can drive progress in the field by clearly defining what constitutes a good explanation and highlighting areas where further research is needed [2]. An important side note is that the terminology within XAI is currently not standardized. For example, "interpretability" and "explainability" are often used interchangeably, which complicates clear communication about XAI concepts [1]. Establishing a standardized, cross-disciplinary terminology would significantly enhance research clarity and effective communication in the field.

Most XAI applications operate in multi-user environments, for instance, multiple doctors may review the same case but require different types of explanations depending on their varying levels of expertise. In summary, collaborative XAI refers to the development of explanations tailored for multi-user scenarios, enabling stakeholders with diverse expertise to understand, interpret and discuss the AI's outputs effectively. To meet these needs, an XAI system must offer multiple views that map technical details into a user-friendly form, provide interactive tools that let users ask follow-up questions and challenge assumptions, thereby fostering a dialogue in which all participants can jointly refine the interpretation. Supporting this perspective, Panigrahi et al. [33] investigated the effects of interactivity on explanations. Their findings demonstrate that interactive strategies such as filtering and overlays enhance users' abilities to comprehend and focus on relevant information. Additionally, the use of counterfactual explanations proved effective in resolving user confusion and clarifying decision boundaries.

Following the discussion on collaborative XAI, the recent rise of Large Language Models (LLMs) presents a promising solution. Their proficiency in natural language generation makes them well-suited for producing explanations accessible to non-expert users. In the literature, the use of LLMs for generating explanations is often referred to as "conversational explanations". These explanations are easier to adapt to individuals since text is more flexible than the static XAI methods [17]. LLMs allow interactive solutions, that put users in charge of customization. One approach are interactive question-answering models, which react to clarifications and questions of users [43]. But this adaptability of LLMs comes with a price in terms of complexity and therefore the required interaction time. On one hand, "augmenting existing XAI methods with conversational user interfaces can increase user engagement and boost user understanding of the AI system" [17]. On the other hand, the generated explanations create an incorrect mental model of the explanation's confidence, leading users to rely too heavily on them [17]. In addition, *LLMs* typically answer verbose, which not only increases the time consumption to read an LLM-generated text [45], but could also lower user satisfaction if they expect a short and concise answer but are confronted with a huge paragraph of explanations. Besides that, we argue that reading an explanation paragraph vastly exceeds the time required to understand a well designed visualization. *LLMs* "generate text using probabilistic models" [19] and "are known to generate inaccurate statements or even hallucinations" [19], which means explanations are not necessarily correct and would again require a verification. Therefore, we propose that AI decisions should not be explained by LLMs, since their decision capability would again need to be verified.

Practical implementation considerations add another layer of complexity. Resource constraints represent a major aspect, as many applications have limited computing power, memory and energy budgets. Therefore, designing explanation methods that strike a balance between explanation quality and computational efficiency is crucial for the widespread adoption of XAI in resource-constrained environments. Current research frequently produces solutions tailored to specific domains, such as medical diagnosis assistance [5, 10, 26, 48] or industrial applications [20, 32], which limits their generalizability. While each domain presents unique requirements, developing more universal approaches would significantly advance the field. Addressing these multifaceted challenges necessitates interdisciplinary collaboration across areas such as algorithm engineering, visualization and social sciences to develop explanations that are both robust and user-friendly. To synthesize the challenges, issues and research gaps outlined above, we have identified the following key research questions:

RQ1 How can human-centered development be integrated into XAI to create contextually appropriate and user-aligned explanations?

RQ2 In what ways can explanations be dynamically adapted to accommodate varying user needs, contextual factors and timing?

RQ3 How can XAI methods be designed to accommodate multi-user settings where stakeholders may have different goals, levels of expertise and requirements for explanations?

RQ4 Which standardized methodologies can be established to systematically evaluate the quality and impact of explanations?

RQ5 How can trust in AI systems be fostered responsibly, including visualizing uncertainty to prevent misplaced trust?

3   Specific Challenges of XAI in Extended Reality

While many of the general challenges of XAI apply across domains, extended reality introduces a distinct set of obstacles due to its immersive, multisensory and highly interactive nature [24]. Unlike traditional 2D interfaces, XR environments blend digital and physical worlds, which requires explanations to be delivered in real time, adapted to rapidly changing user contexts and integrated seamlessly into complex spatial and sensory experiences in order to not distract users [14]. These unique characteristics amplify existing XAI challenges, such as cognitive overload, personalization and timing, while also introducing new requirements for multimodal data integration, spatial anchoring of explanations and collaborative multi-user scenarios. In this section, we analyze how the core issues of explainability manifest in XR, highlighting the need for novel approaches that go beyond conventional XAI methods.

A significant concern of Explainable AI in extended reality is the risk of cognitive overload: users immersed in XR environments already process a high density of sensory and interactive information and additional explanations may inadvertently increase cognitive demand, potentially overwhelming users. The integration of XAI visualizations therefore requires a careful design that contributes positively to the user experience rather than detracting from it. Yenduri et al. [14] found that disruptions or delays of the experience caused by XAI systems can negatively affect user engagement. This finding is particularly pronounced in fully virtual environments, but it also holds relevance for general XAI approaches. As for data visualizations in XR, XAI visualizations also need to be designed carefully, to not occlude important parts of the digital environment, break the immersion or lead to user distraction. To reduce overload on the visual channel, research explored haptic and auditory explanations as a promising alternative channel [40, 46]. For that, vibration motors in the headset or controllers, as well as the device's speakers, can be utilized. Since XR devices must be designed in a compact and lightweight form to ensure comfort during all-day use, their computational power is often more limited than that of desktops or handheld devices such as tablets. As a result, many XR applications rely on cloud-based computing solutions to handle more demanding processing tasks. Nevertheless, explanations are often required in real time, which means that the latency introduced by cloud-based systems must be carefully considered when designing visualizations.

Xu et al. [46] provide an in-depth analysis of integrating XAI into XR. They highlight that XR systems significantly increase the complexity of generating relevant and context-aware explanations, by capturing rich real-time sensory data about users and their environments. In particular, "AR has a much deeper real-time understanding of a user's current state via the sensors within an HMD [than normal desktop PCs or handhelds]" [46], which underscores the necessity for dynamically personalized explanations that can adapt to rapidly changing user and environmental conditions. Furthermore, the explanation must be contextually relevant and comprehensible, bridging the gap between complex AI processes and a user's intuitive understanding of the visible scenario.

When discussing XR technologies, one cannot overlook one of the most significant developments: the Metaverse [28], an interconnected, multi-user XR environment. Yenduri et al. [14] explored the integration of Explainable AI within the Metaverse. The authors' findings align with those of Xu et al. [46] that the vast amount of data generated during XR interactions, combined with the challenges of providing explanations in dynamic, real-time environments, makes integrating XAI visualizations significantly more complex than in traditional 2D interfaces. Additionally, explanations in XR must often be available on demand and spatially anchored within the physical or virtual environment. This further complicates decisions about when and how to present explanations, as it is well established that "explanations need to be tailored to a user's state and context" [46]. The data-collection process in XR is multifaceted, involving eye trackers, motion detectors, biometric devices and other sensors, which requires robust methods that articulate how this complex data informs an AI's decisions [46]. Another prominent example of multi-user extended reality is the *remote-expert scenario*. In this setup, a remote specialist provides real-time guidance to an on-site user through the use of immersive technologies. The expert observes a user's environment through live video streams or spatial data and delivers instructions using annotations, holograms and/or avatars. This approach enables efficient remote collaboration, especially in fields such as maintenance, healthcare and education. For instance, Chang et al. [6] examine the effectiveness of virtual replicas in enhancing remote-expert collaboration in their paper "Efficient VR-AR communication method using virtual replicas in XR remote collaboration." Similarly, Oppermann et al. [32] investigate the concept of an "Industrial Metaverse" for remote maintenance using avatars and digital twins. When integrating AI and therefore Explainable AI into such frameworks, a key question arises: How should explanations be delivered to collaborators with varying expertise? There are two primary strategies: (1) Tailoring the explanations to each expertise level, which may hinder collaborative discussion if participants receive different information or (2) Providing uniform explanations to all collaborators, which can facilitate collaboration but risks being too simplistic for experts or too complex for novices. In addition to the expertise factor, the technology used to experience XR content, such as head-mounted (HMD) and hand-held (HHD) displays, CAVEs and Powerwalls, adds another layer of complexity. These devices impose different spatial, interactive and design constraints that strongly influence how XAI can be presented. Devices such as HMDs and HHDs offer individual spaces in which explanations can be tailored to a single user, even during collaboration. In contrast, shared display environments such as CAVEs or Powerwalls provide one common visualization, requiring explanations that serve multiple viewers simultaneously. Therefore, XAI in XR must adapt to both device-specific capabilities and the different environments.

The absence of standardized approaches in general XAI research also extends to its application in XR environments. Maathuis et al. highlight "there is a need for more standardized evaluation frameworks (including benchmarks and metrics) that can effectively assess both technical accuracy and human interpretability (the quality of explanations)" [24]. This includes considering factors like immersion, presence, task performance and cognitive load, which are particularly relevant in XR contexts. The adoption of such standards can enable automatic validation, thereby enhancing the quality of the provided explanations.

Besides the general challenges presented in Chapter 2, the XR environment leads to additional and more specific research questions:

XRQ1 How can multisensory XR data be integrated into XAI to support a more comprehensive understanding of AI behavior?

XRQ2 How can explanations be delivered in real-time and adapted to a user's rapidly changing context, actions and virtual environment without causing disruption or cognitive overload?

XRQ3 How can XAI be applied in multi-user XR environments (e.g., different spaces and devices) to generate explanations that are understandable to all participants, considering their diverse perspectives and varying levels of expertise?

XRQ4 What are appropriate evaluation metrics for assessing the effectiveness and usability of XAI methods in XR, particularly with respect to immersion, presence and cognitive load?

4   Conclusion

Current approaches to XAI remain insufficient for meeting the complex, context-rich demands posed by AI-enhanced XR applications. As these systems increasingly permeate safety-critical and socially sensitive domains, such as healthcare and education, the need for transparent, adaptive and user-centered explanatory mechanisms becomes even more urgent. Static, one-size-fits-all solutions are inadequate in these dynamic, multi-user environments.

This review outlines four central research priorities in the field of XAI, with particular relevance to both general applications (RQ1-5) and those situated within XR contexts (XRQ1-4). These priorities are considered essential for advancing the effectiveness, interpretability and real-world applicability of XAI methods. The four key areas of focus are as follows:

- Integration of human-centered design principles into XAI systems,
- Dynamic adaptation of explanations to users, context, goals,
- Development of robust and standardized evaluation frameworks and
- Integration of responsible trust through uncertainty-aware explanations.

Within XR environments, these challenges are further amplified by multisensory data streams, real-time interaction demands, evolving user and environmental states and heterogeneous levels of user expertise. Effectively addressing these challenges requires the development of novel, context-sensitive explanatory strategies that are both multimodal and scalable, while avoiding cognitive overload. Meeting these challenges is essential to ensuring that AI-augmented XR systems are not only technically robust but also socially responsible and genuinely aligned with human needs and values.

Acknowledgment. We gratefully acknowledge the funding of the research training group VAMoS provided by the Ministry of Science and Health of the German federal state of Rhineland-Palatinate. Moreover, I would like to thank Kai Felix Mannweiler for the valuable discussions and for his support in preparing this paper.

Disclosure of Interests. The authors have no competing interests to declare that are relevant to the content of this article.

References

1. Barredo Arrieta, A., Díaz-Rodríguez, N., et al., J.D.: Explainable artificial intelligence (XAI): concepts, taxonomies, opportunities and challenges toward responsible AI. Inf. Fus. **58**, 82–115 (2020). https://doi.org/10.1016/j.inffus.2019.12.012
2. Bhattacharya, A., Verbert, K.: how good is your explanation?: towards a standardised evaluation approach for diverse XAI methods on multiple dimensions of explainability. In: Adjunct Proceedings of the 32nd ACM Conference UMAP, pp. 513–515. UMAP Adjunct 2024, ACM, New York, NY, USA (2024). https://doi.org/10.1145/3631700.3664911
3. Bilal, A., Ebert, D., Lin, B.: LLMS for explainable AI: a comprehensive survey (2025). https://doi.org/10.48550/arXiv.2504.00125
4. Brunotte, W., Chazette, L., Klös, V., Speith, T.: Quo vadis, explainability? – a research roadmap for explainability engineering. In: Requirements Engineering: Foundation for Software Quality: 28th International Working Conference, REFSQ 2022, Birmingham, UK, March 21–24, 2022, Proceedings, pp. 26–32. Springer-Verlag, Berlin, Heidelberg (2022). https://doi.org/10.1007/978-3-030-98464-9_3
5. Calisto, F.M., Fernandes, J., Morais, M.: Assertiveness-based agent communication for a personalized medicine on medical imaging diagnosis. In: Proceedings of the 2023 CHI Conference. CHI 2023, ACM, New York, NY, USA (2023). https://doi.org/10.1145/3544548.3580682
6. Chang, E., Lee, Y., Billinghurst, M., Yoo, B.: Efficient VR-AR communication method using virtual replicas in XR remote collaboration. Int. J. Hum Comput Stud. **190**, 103304 (2024). https://doi.org/10.1016/j.ijhcs.2024.103304
7. Chattopadhay, A., Sarkar, A., Howlader, P.: Grad-CAM++: generalized gradient-based visual explanations for deep convolutional networks. In: 2018 IEEE Winter Conference on Applications of Computer Vision (WACV), pp. 839–847. IEEE, New York, NY, USA (2018). https://doi.org/10.1109/WACV.2018.0009
8. Chen, C., Liao, M., Sundar, S.S.: When to explain? Exploring the effects of explanation timing on user perceptions and trust in ai systems. In: Proceedings of the Second International Symposium on Trustworthy Autonomous Systems. TAS 2024, ACM, New York, NY, USA (2024). https://doi.org/10.1145/3686038.3686066
9. Chromik, M., Eiband, M., Völkel, S.T., Buschek, D.: Dark patterns of explainability, transparency, and user control for intelligent systems. In: IUI Workshops, vol. 2327, (2019)

10. Corti, L., Oltmans, R., Jung, J.: it is a moving process: understanding the evolution of explainability needs of clinicians in pulmonary medicine. In: Proceedings of the 2024 CHI Conference, ACM, New York, NY, USA (2024). https://doi.org/10.1145/3613904.3642551
11. Dwivedi, R., D.: Explainable AI (XAI): Core ideas, techniques, and solutions. ACM Comput. Surv. 55(9), (2023). https://doi.org/10.1145/3561048
12. European Union: Regulation (eu) 2016/679 of the European parliament and of the council of 27 April 2016 on the protection of natural persons with regard to the processing of personal data and on the free movement of such data, and repealing directive 95/46/ec (general data protection regulation) (2016). http://data.europa.eu/eli/reg/2016/679/oj
13. European Union: Regulation (eu) 2024/1689 of the european parliament and of the council of 13 June 2024 on laying down harmonised rules on artificial intelligence and amending regulations (ec) no 300/2008, (eu) no 167/2013, (eu) no 168/2013, (eu) 2018/858, (eu) 2018/1139 and (eu) 2019/2144 and directives 2014/90/eu, (eu) 2016/1797 and (eu) 2018/1808 and repealing regulation (eu) 2022/2042 (2024). http://data.europa.eu/eli/reg/2024/1689/oj
14. G, C.S., Yenduri, G., Srivastava, G.: Explainable ai for the metaverse: a short survey. In: 2023 International Conference on Intelligent Metaverse Technologies & Applications (iMETA), pp. 1–6. IEEE, New York, NY, USA (2023). https://doi.org/10.1109/iMETA59369.2023.10294907
15. Greenes, R.: Clinical Decision Support: The Road to Broad Adoption, Elsevier Inc (2014). https://doi.org/10.1016/C2012-0-00304-3. 2nd edn
16. Guidotti, R., Monreale, A., Ruggieri, S.: A survey of methods for explaining black box models. ACM Comput. Surv. 51(5), (2018). https://doi.org/10.1145/3236009
17. He, G., Aishwarya, N., Gadiraju, U.: Is conversational XAI all you need? Human-AI decision making with a conversational XAI assistant. In: Proceedings of the 30th International Conference on Intelligent User Interfaces, pp. 907–924. ACM, New York, NY, USA (2025). https://doi.org/10.1145/3708359.3712133 IUI 2025
18. Kim, Y., Aamir, Z., Singh, M.: Explainable XR: understanding user behaviors of XR environments using LLM-assisted analytics framework. IEEE Trans. Visual. Comput. Graph. 31(5), 2756–2766 (2025). https://doi.org/10.1109/TVCG.2025.3549537
19. Kosasih, E.E., Papadakis, E., Baryannis, G., Brintrup, A.: A review of explainable artificial intelligence in supply chain management using neurosymbolic approaches. Int. J. Prod. Res. 62(4), 1510–1540 (2024)
20. Kostopoulos, G., Davrazos, G., Kotsiantis, S.: Explainable artificial intelligence-based decision support systems: a recent review. Electronics 13(14), (2024). https://doi.org/10.3390/electronics13142842
21. Kotte, H., Daiber, F., Kravcik, M.: FitSight: tracking and feedback engine for personalized fitness training. In: Proceedings of the 32nd ACM Conference on User Modeling, Adaptation and Personalization, pp. 223–231. UMAP 2024, ACM, New York, NY, USA (2024). https://doi.org/10.1145/3627043.3659547
22. Kästner, L., Langer, M., Lazar, V.: On the relation of trust and explainability: why to engineer for trustworthiness. In: 2021 IEEE 29th International Requirements Engineering Conference Workshops (REW), pp. 169–175. IEEE, New York, NY, USA (2021). https://doi.org/10.1109/REW53955.2021.00031
23. Lundberg, S.M., Lee, S.I.: A unified approach to interpreting model predictions. In: Proceedings of the 31st International Conference on Neural Information Processing Systems, pp. 4768–4777. NIPS 2017, Curran Associates Inc., Red Hook, NY, USA (2017)

24. Maathuis, C., Cidota, M.A., Datcu, D., Marin, L.: Integrating explainable artificial intelligence in extended reality environments: a systematic survey. Mathematics 13(2), (2025). https://doi.org/10.3390/math13020290
25. Manger, C.: Explainability in automated parking: The effect of augmented reality visualizations on user experience and situation awareness. In: Proceedings of the 22nd International Conference on Mobile and Ubiquitous Multimedia, pp. 152–158. Association for Computing Machinery, New York, NY, USA (2023). https://doi.org/10.1145/3626705.3627796 MUM 2023
26. Markus, A.F., Kors, J.A., Rijnbeek, P.R.: The role of explainability in creating trustworthy artificial intelligence for health care: a comprehensive survey of the terminology, design choices, and evaluation strategies. J. Biomed. Inform. **113**, 103655 (2021). https://doi.org/10.1016/j.jbi.2020.103655
27. Langer, M., et al.: What do we want from explainable artificial intelligence (XAI)? - a stakeholder perspective on XAI and a conceptual model guiding interdisciplinary XAI research. Artif. Intell. **296**, 103473 (2021). https://doi.org/10.1016/j.artint.2021.103473
28. Meta: Metaverse (2025). https://www.meta.com/de-de/metaverse/
29. Miller, T.: Explanation in artificial intelligence: insights from the social sciences. Artif. Intell. **267**, 1–38 (2019). https://doi.org/10.1016/j.artint.2018.07.007
30. Mohammed, A.A.A.: Adaptive explainable AI: personalizing machine explanations based on user expertise levels. J. Posthumanism **5**(7), 317–334 (2025). https://doi.org/10.63332/joph.v5i7.2793
31. Nimmy, S.F., Hussain, O.K., Chakrabortty, R.K., Hussain, F.K., Saberi, M.: Explainability in supply chain operational risk management: a systematic literature review. Knowl. Based Syst. **235**, 107587 (2022). https://doi.org/10.1016/j.knosys.2021.107587
32. Oppermann, L., Buchholz, F., Uzun, Y.: Industrial metaverse: supporting remote maintenance with avatars and digital twins in collaborative XR environments. In: Extended Abstracts of the 2023 CHI Conference. CHI EA 2023, ACM, New York, NY, USA (2023). https://doi.org/10.1145/3544549.3585835
33. Panigrahi, I., Kim, S.S.Y., Liaqat, A.: Interactivity X explainability: toward understanding how interactivity can improve computer vision explanations. In: Proceedings of the Extended Abstracts of the CHI Conference. CHI EA 2025, ACM, New York, NY, USA (2025). https://doi.org/10.1145/3706599.3719730
34. Poché, A., Hervier, L., Bakkay, M.C.: Natural example-based explainability: a survey (2023). https://doi.org/10.48550/arXiv.2309.03234
35. Poyiadzi, R., Sokol, K., Santos-Rodriguez, R.: Face: Feasible and actionable counterfactual explanations. In: Proceedings of the AAAI/ACM Conference on AI, Ethics, and Society, p. 344–350. AIES 2020, ACM, New York, NY, USA (2020). https://doi.org/10.1145/3375627.3375850
36. Prabhudesai, S., Yang, L., Asthana, S.: Understanding uncertainty: how lay decision-makers perceive and interpret uncertainty in human-AI decision making. In: Proceedings of the 28th International Conference on Intelligent User Interfaces, pp. 379–396. ACM, New York, NY, USA (2023). https://doi.org/10.1145/3581641.3584033 IUI 2023
37. Ribeiro, M.T., Singh, S., Guestrin, C.: Anchors: High-precision model-agnostic explanations. In: Proceedings of the AAAI Conference on Artificial Intelligence, vol. 32, no. 1 (2018). https://doi.org/10.1609/aaai.v32i1.11491

38. Ribeiro, M.T., Singh, S., Guestrin, C.: why should i trust you?: explaining the predictions of any classifier. In: Proceedings of the 22nd ACM SIGKDD International Conference on Knowledge Discovery and Data Mining, pp. 1135–1144. KDD 2016, ACM, New York, NY, USA (2016). https://doi.org/10.1145/2939672.2939778
39. Ryu, M., Cha, S.H.: Context-awareness based driving assistance system for autonomous vehicles. Int. J. Control Autom. **11**(1), 153–162 (2018)
40. Schuller, B.W., Virtanen, T., Riveiro, M.: Towards sonification in multimodal and user-friendly explainable artificial intelligence. In: Proceedings of the 2021 ICMI, pp. 788–792. ACM (2021). https://doi.org/10.1145/3462244.3479879
41. Selvaraju, R.R., Cogswell, M., Das, A.: Grad-cam: visual explanations from deep networks via gradient-based localization. IEEE ICCV , 618–626 (2017). https://doi.org/10.1109/ICCV.2017.74
42. Shajalal, M., Boden, A.: Explaining ai decisions: towards achieving human-centered explainability in smart home environments. In: Explainable Artificial Intelligence, pp. 418–440. Springer Nature Switzerland, Cham (2024). https://doi.org/10.1007/978-3-031-63803-9_23
43. Slack, D., Krishna, S., Lakkaraju, H.: Explaining machine learning models with interactive natural language conversations using TalkToModel. Nat. Mach. Intell. **5**(8), 873–883 (2023). https://doi.org/10.1038/s42256-023-00692-8
44. Vainio-Pekka, H., Agbese, M.O.O., Jantunen, M.: The role of explainable ai in the research field of ai ethics. ACM Trans. Interact. Intell. Syst 13(4), (2023). https://doi.org/10.1145/3599974
45. Wang, X., Yu, M.: Less or more: towards glanceable explanations for LLM recommendations using ultra-small devices. In: Proceedings of the 30th International Conference on Intelligent User Interfaces, pp. 938–951. ACM, New York, NY, USA (2025). https://doi.org/10.1145/3708359.3712074 IUI 2025
46. Xu, X., Yu, A., Jonker, T.R.: XAIR: a framework of explainable ai in augmented reality. In: Proceedings of the 2023 CHI Conference. CHI 2023, ACM, New York, NY, USA (2023). https://doi.org/10.1145/3544548.3581500
47. Yang, L.H., Liu, J., et al., F.F.Y.: Highly explainable cumulative belief rule-based system with effective rule-base modeling and inference scheme. Knowl. Based Syst. **240**, 107805 (2022). https://doi.org/10.1016/j.knosys.2021.107805
48. Yang, Q., Hao, Y., Quan, K.: Harnessing biomedical literature to calibrate clinicians' trust in ai decision support systems. In: Proceedings of the 2023 CHI Conference. CHI 2023, ACM, New York, NY, USA (2023). https://doi.org/10.1145/3544548.3581393
49. Zafar, M.R., Khan, N.: Deterministic local interpretable model-agnostic explanations for stable explainability. Mach. Learn. Knowl. Extract. **3**(3), 525–541 (2021). https://doi.org/10.3390/make3030027
50. Zhou, J., Chen, F., Holzinger, A.: Towards explainability for AI Fairness, pp. 375–386. Springer International Publishing (2022). https://doi.org/10.1007/978-3-031-04083-2_18

Silent Interfaces for Situation Awareness in Active Assisted Living: A Human-Centered Approach to Proactive Assistance

Peter Klein[1]([✉]) [iD] and Reiner Wichert[2] [iD]

[1] uCORE Systems GmbH, Andernach, Germany
peter.klein@ucore-systems.com
[2] Hochschule Darmstadt, Darmstadt, Germany
reiner.wichert@h-da.de
https://www.ucore-systems.com , https://www.h-da.de

Abstract. As Active Assisted Living (AAL) systems evolve, a key challenge is to provide timely assistance without overwhelming users with constant interaction. We introduce the concept of Silent Interfaces—interaction paradigms that remain unobtrusive in everyday life yet engage precisely when necessary. Building on principles from Calm Technology and Situation Awareness, we outline a human-centered design framework and a technical architecture based on semantic rule sets, sensor fusion, and AI-driven event interpretation.

Our approach, implemented in the uCORE platform, focuses on multimodal, context-triggered cues that range from subtle visual or haptic signals to autonomous emergency calls. Field deployments in assisted living facilities indicate increased user acceptance, reduced alert fatigue, and improved safety, without compromising privacy or autonomy. While these insights stem from practical operation rather than controlled studies, they highlight the potential of Silent Interfaces to extend human perception and enable proactive care. The concept offers guidance for designing unobtrusive, trustworthy, and scalable AAL solutions.

Keywords: Silent Interfaces · Situation Awareness · Active Assisted Living · Calm Technology · Human-Centered Design · Ethical Design

1 Introduction

Demographic change and the rising demand for long-term care have led to increased interest in Active Assisted Living (AAL) systems that help older adults and people with special needs live independently and safely. Advances in smart home technology, sensor fusion, and artificial intelligence now enable highly capable assistance systems, yet adoption remains limited due to concerns about complexity, intrusiveness, and loss of autonomy [4].

Many existing AAL solutions rely on frequent user interactions or visible monitoring devices, which can create interaction fatigue and reduce acceptance. Those AAL systems often become obtrusive not only through frequent notifications, but also through intrusive monitoring practices. For example, systems that rely on persistent camera surveillance, mandatory wearable confirmations, or recurring audio prompts can disrupt daily routines, heighten feelings of being monitored, and lead to long-term alert fatigue. These scenarios illustrate how current solutions may unintentionally undermine autonomy and comfort—ultimately motivating the need for more unobtrusive, context-sensitive interaction paradigms such as Silent Interfaces.

This raises a key design question: How can we provide reliable, context-aware assistance without burdening users with constant notifications, visual clutter, or manual operations?

We introduce the concept of Silent Interfaces as a human-centered response to this challenge. These interfaces remain unobtrusive during normal operation, blending into the environment, yet become precisely visible when a relevant situation arises. By combining principles from Calm Technology [18] and Situation Awareness [1,6] with AI-driven event interpretation, Silent Interfaces act as an extension of human perception—alerting users only when meaningful action is required.

Our work builds on more than two years of iterative design and deployment in assisted living facilities, using the uCORE platform as the technical foundation.[1] The platform integrates semantic rule sets and sensor fusion to achieve highly precise situation detection, informing residents of the detected situation via multimodal feedback and including autonomous emergency calls certified with zero false alarms. This involves users in the decision-making process, which leads to increased acceptance. We aim to show that AAL systems can be both proactive and unobtrusive, offering a scalable model for care-oriented environments.

2 Related Work

The design of unobtrusive assistance systems draws on established research in Human–Computer Interaction (HCI), ubiquitous computing, and assistive technologies. This section reviews three key areas forming the conceptual and technical foundation for Silent Interfaces: Calm Technology, Situation Awareness, and AI in Active Assisted Living.

2.1 Calm Technology and Ambient User Experience

The concept of Calm Technology, introduced by Weiser and Brown [17,18], advocates for systems that remain in the user's periphery of attention and move to the center only when necessary. This reduces cognitive load and supports long-term

---

[1] uCORE builds upon and extends concepts from the Fraunhofer-led universAAL initiative, reimplementing semantic reasoning and service-oriented integration with updated AI components [7].

acceptance. Recent work on Ambient User Experience extends these principles to multi-sensory, context-aware cues [5], informing the design of Silent Interfaces.

2.2 Situation Awareness Models

Situation Awareness (SA), as defined by Endsley [6], describes the perception, comprehension, and projection of environmental elements. While originally developed for aviation and other high-risk domains, it has been adapted for healthcare and AAL [1]. In these contexts, SA provides a framework for determining when and how systems should engage users.

2.3 AI in Active Assisted Living

AI in AAL leverages techniques such as sensor fusion, semantic reasoning, and anomaly detection to improve precision in context recognition [4]. Semantic, ontology-driven approaches, as seen in platforms like universAAL [7], enable structured interpretation of heterogeneous data. Combining these methods with human-centered interaction models helps address challenges such as reducing false alarms and maintaining user trust.

In contrast, machine learning typically depends on large datasets to identify behavioral patterns and recognize situations. Yet, because human behavior is highly individual, generic models may fall short in capturing personal nuances. This calls for a stronger focus on individual-level data. However, the available personal data alone is often insufficient to reliably determine a user's current context, making direct user involvement necessary to confirm and validate recognized situations [2].

3   Silent Interfaces: Concept and Design Principles

3.1 Definition and Characteristics

We define a Silent Interface as an interaction paradigm that operates unobtrusively during normal conditions, yet becomes salient and actionable when a situation of relevance arises. Unlike conventional interfaces, which often rely on continuous user attention, Silent Interfaces remain in the periphery of perception until activation is warranted by contextual changes. This minimizes cognitive load, avoids interaction fatigue, and supports long-term acceptance.

A Silent Interface is fundamentally defined by four core characteristics that distinguish it from both conventional and fully invisible interaction paradigms.

First, it is context-aware, meaning it continuously interprets environmental and physiological data to assess the relevance of a situation. This involves the real-time analysis of sensor inputs such as motion, temperature, proximity, or biometric signals. By semantically interpreting these heterogeneous data streams, the system determines when a specific condition or anomaly requires the user's attention—ensuring that the interface remains dormant during non-critical states and becomes responsive only when genuinely needed.

Second, Silent Interfaces are designed with low interaction demand. Unlike traditional systems that rely on frequent prompts, confirmations, or manual controls, Silent Interfaces minimize the need for active user engagement during everyday operation. The system functions autonomously in the background, relieving users from constant decision-making or attention, which is particularly valuable for individuals with cognitive, physical, or attentional limitations.

Third, the interfaces offer multimodal signalling tailored to user preferences and accessibility needs. Depending on the user's context and capabilities, feedback can be delivered via different ambient light changes, auditory cues, or a combination thereof. The modality and form of these signals are customizable and inclusive, adhering to universal design principles to support a diverse range of users—whether elderly, visually impaired, or non-native language speakers.

Fourth, Silent Interfaces are governed by a principle of scalable salience. The intensity, modality, and intrusiveness of feedback dynamically adjust based on the urgency and nature of the detected event. For example, a mild, non-urgent context might trigger a soft ambient glow with green LEDs, whereas a critical situation—such as a fall or fire risk—could escalate to direct voice prompts or autonomous emergency calls with an ambient glow by red LEDs. This scalability ensures that the interface communicates with appropriate gravity, avoiding over-alerting while ensuring action when truly necessary. Together, these characteristics form a foundation for building assistance systems that are proactive yet unobtrusive—technologically capable without becoming socially or cognitively invasive.

3.2   Difference to Invisible Interfaces

While both Silent and Invisible Interfaces aim to reduce visual clutter and minimize unnecessary user interaction, they differ significantly in their design philosophy and degree of user involvement. Invisible Interfaces, as envisioned in the paradigm of ubiquitous computing, strive for complete technological concealment. They often operate autonomously in the background, executing tasks without explicit user input or even awareness. The goal is seamless integration into daily life, where interaction becomes implicit or is entirely abstracted away.

Silent Interfaces, by contrast, preserve a latent communicative channel that remains inactive until user awareness or intervention becomes necessary. Rather than fully automating all processes, Silent Interfaces are designed to stay unobtrusive while retaining the capacity to engage the user meaningfully when required. This aligns with the principle of keeping the user "in the loop," supporting situational awareness and control without overwhelming the user with constant feedback or notifications.

In the context of AAL systems, this distinction is particularly relevant: while Invisible Interfaces may support convenience through full automation, Silent Interfaces provide a more balanced approach—respecting user autonomy, ensuring transparency, and enabling trust in sensitive environments such as healthcare or assisted living.

3.3 Design Principles

Building on the concepts of Calm Technology and Situation Awareness, we define and operationalize six core design principles that guide the development and evaluation of our system. Each principle is grounded in human-centered interaction design and is paired with measurable indicators:

- **Peripheral Presence:** The system maintains a non-intrusive presence in the user's environment, remaining largely in the background unless needed. This is evaluated through user-reported perceived intrusiveness [8].
- **Context-Triggered Engagement:** Interaction is initiated only in relevant situations, based on contextual cues. This principle is assessed via trigger precision/recall metrics and false positive rates, reflecting the system's ability to engage at appropriate times.
- **Progressive Disclosure:** Information and interaction unfold in stages, revealing only what is necessary at a given moment. This is measured through escalation compliance curves and user-rated appropriateness of notifications or prompts.
- **Multimodal Accessibility:** The system supports diverse input and output modalities, ensuring accessibility across user abilities and preferences. Evaluation is based on compliance with recognized standards such as WCAG 2.1 [15].
- **User-Centric Adaptation:** Behavior adapts to user preferences and feedback over time, aiming to reduce friction and increase satisfaction. This is measured through reductions in nuisance notifications and improvements in user satisfaction scores [12].
- **Transparency of Action:** Users are informed about the system's decisions and actions in an understandable way. This principle is evaluated through explanation satisfaction metrics [10], reflecting clarity and perceived usefulness of system feedback.

4 Technical Implementation

The Silent Interfaces described in this paper are implemented within the uCORE platform, a modular infrastructure that integrates hardware, operating system, and application layers to enable high-precision situation detection in Active Assisted Living (AAL) environments. The platform comprises three core components: CORE ONE (smart home controller hardware), CORE OS (semantic operating system), and CORE ACCESS (user-facing applications for various stakeholders).

To support system-level clarity, we provide a brief overview of how the components described in this section interact during normal operation. In a typical deployment, the system continuously gathers data from distributed ambient sensors, processes and interprets this data locally on CORE ONE, and triggers context-aware, low-salience output through ambient lighting, voice prompts, or caregiver dashboards. This end-to-end process ensures that Silent Interfaces remain both unobtrusive and operationally coherent in real-world scenarios.

4.1 System Architecture

The architecture adopts a layered approach inspired by open AAL platforms such as universAAL [7].

To provide a clearer understanding of the system's operation in real-world deployments, we briefly outline the typical hardware environment. A standard installation includes CORE ONE as the local processing unit, non-invasive environmental sensors (motion, temperature, humidity, CO_2), optional physiological sensors, and ambient output devices such as LED modules or wall-mounted displays. These components form the physical substrate through which the layered architecture is realized.

1. **Sensing Layer:** A heterogeneous network of environmental sensors (temperature, humidity, motion, light), physiological sensors (heart rate, respiration), and utility meters (energy, water) captures multimodal data streams.
2. **Processing Layer:** Data is processed locally on the CORE ONE controller running CORE OS, which performs sensor fusion and initial preprocessing to filter noise and detect relevant patterns in near real time.
3. **Semantic Reasoning Layer:** Leveraging ontology- and service-oriented approaches, CORE OS employs a domain-specific semantic model to contextually interpret events. Rules correlate heterogeneous sensor signals with user routines, predefined thresholds, and safety-relevant conditions.
4. **Application Layer:** The CORE ACCESS module delivers role-specific interfaces for various stakeholders (residents, caregivers, facility managers, emergency services), facilitating interaction and oversight.

In practice, these layers operate together as a continuous data-flow pipeline: sensors collect environmental and physiological signals, CORE ONE preprocesses and aggregates them, the semantic model interprets their contextual meaning, and CORE ACCESS provides the corresponding low-salience feedback to residents or caregivers. This high-level operational chain offers a holistic view of how Silent Interfaces function end-to-end in deployment environments.

4.2 AI-Based Event Interpretation

To achieve precise and trustworthy situation awareness with minimal false positives, the system employs a hybrid approach:

- **Rule-based reasoning:** Encodes expert knowledge and formalized safety regulations for deterministic handling of well-defined conditions (e.g., fall detection thresholds, prolonged absence from bed).
- **Machine learning models (BETA):** Supplementary models under evaluation are used to detect non-explicit patterns and anomalies not captured by the rule base. These models are currently not deployed in safety-critical core functions.

– **Human-in-the-Loop (HITL) confirmation:** Rather than relying solely on probabilistic confidence levels, the system integrates user or caregiver feedback for interpretive validation. In case of a potential emergency, the resident is prompted to confirm or dismiss the event. Lack of response within a context-sensitive timeframe initiates an escalation protocol toward caregivers or emergency services [2].

4.3 Multimodal Feedback Strategies

The uCORE platform implements a graded, context-aware feedback model, enabling silent operation while ensuring timely escalation when necessary:

1. **Subtle cues:** Visual indicators (ambient light changes), haptic signals (wearable vibration), or soft audio tones for non-critical notifications.
2. **Escalated cues:** Persistent or worsening situations escalate to more salient modalities, such as flashing lights or direct voice prompts.
3. **Critical alerts:** In high-confidence or emergency situations, the system autonomously initiates emergency calls, optionally enriched with contextual data in real time (e.g., in case of a fire alarm, notifying first responders whether and where people are present in the building).

5 Human-Centered Design Approach

The development and implementation of Silent Interfaces within the uCORE platform followed a human-centered design (HCD) process based on ISO 9241–210 [11], ensuring that technical capabilities, such as sensor precision and semantic context-awareness, were meaningfully translated into real-world usability, acceptance, and trust among all stakeholders in Active Assisted Living (AAL) environments.

5.1 Scenario-Based Design for Unseen Interaction

Given the unobtrusive nature of Silent Interfaces, conventional interface prototyping methods proved insufficient. We therefore adapted Scenario-Based Design (SBD) [13] to model interaction patterns that are not visually apparent. These scenarios captured typical daily routines, potential critical events such as falls or prolonged inactivity, and edge cases where system interpretation might be ambiguous. Iterative refinement through field observations, interviews, and participatory design workshops helped align activation thresholds and modalities with user expectations and comfort levels.

5.2 Stakeholder Involvement

AAL environments involve diverse stakeholders with varying needs, technical literacy, and operational priorities. Our process engaged residents seeking comfort and privacy, caregivers needing timely and actionable alerts, relatives wanting reassurance without overload, facility managers interested in efficiency, and emergency services requiring high-confidence notifications. This diversity often revealed conflicting requirements, which were addressed through configurable interaction modalities rather than one-size-fits-all solutions.

5.3 Iterative Prototyping and Field Testing

Prototyping cycles combined laboratory simulation with deployment in real-world environments, most notably the WoQuaZ living lab in Weiterstadt [9,14]. Documented as part of large-scale AAL initiatives such as ACTIVAGE, WoQuaZ provided a realistic testbed for validating detection algorithms, assessing user reactions to different escalation patterns, and identifying usability gaps such as unnoticed subtle cues. Findings from each cycle informed both the semantic rule set in CORE OS and the interaction logic in CORE ACCESS. Each iteration was benchmarked against ISO 9241–11 usability criteria (effectiveness, efficiency, satisfaction).

5.4 Ethical and Privacy Considerations

Ethical design was supported by established frameworks such as MEESTAR [16] and the EVA (Ethical Value Assessment) model [3]. EVA operationalizes ethical principles by linking them to concrete product requirements, stakeholder priorities, and business considerations. In line with these frameworks, Silent Interfaces were designed to operate without continuous audiovisual recording, relying instead on privacy-preserving sensor modalities and on-device processing to avoid unnecessary data collection.

6 Use Cases

The following use cases illustrate how Silent Interfaces, implemented within the uCORE platform, translate the principles of unobtrusive yet context-relevant interaction into tangible benefits for residents, caregivers, and facility operators in Active Assisted Living environments.

6.1 Acceptance Indicators

During pilot operations in assisted living facilities, feedback from residents, caregivers, and other stakeholders consistently indicated that the system was perceived as unobtrusive in daily life and that alerts were generally regarded as relevant and trustworthy. Stakeholders also described their overall experience

with the system in positive terms, citing reduced disruption and better alignment with real care needs.

User feedback was collected through structured informal interviews ($N = 27$)
with residents, caregivers, and facility staff after each system iteration. Data was
thematically analyzed using an open coding approach, revealing recurring patterns of high acceptance, particularly in relation to perceived unobtrusiveness,
relevance of alerts, and reduction of "alarm fatigue." While not statistically representative, the consistency across three distinct deployment sites supports the
internal validity of these findings.

6.2   Comfort: Energy Efficiency via Subtle Ventilation Cues

Beyond safety, Silent Interfaces also address comfort and sustainability. For
example, indoor air quality is monitored alongside weather forecasts, energy
prices, and building temperature. Poor indoor air quality (elevated CO_2 levels,
humidity) is a common issue in assisted living environments, negatively impacting comfort and cognitive performance. Manual ventilation is often neglected,
especially among elderly residents with reduced sensory awareness or mobility
constraints.

Here the Silent Interface response is implemented within the uCORE platform by continuously monitoring indoor air quality via non-invasive sensors
and contextual data. Upon detecting suboptimal ventilation conditions during periods conducive to airing (e.g., mild temperatures), the system provides
subtle feedback through ambient lighting changes or via speech. This proactive, non-intrusive approach encourages residents to ventilate at optimal times
without imposing mandatory actions, resulting in measurable energy savings
and improved air quality. These use cases demonstrate how context-aware,
low-salience interaction can promote sustainable and health-enhancing behavior without relying on overt instructions or alarms. It supports the principle of
Peripheral Presence and illustrates a practical implementation of Calm Technology in daily routines.

6.3   Operational: Care Coordination Without Additional Devices

In multi-resident facilities, caregivers often rely on disparate communication
channels, leading to inefficiencies and missed opportunities for early intervention. CORE ACCESS integrates with existing workflows, providing consolidated
situation updates on mobile devices or wall-mounted displays in staff areas. For
example, if a resident's routine deviates significantly—such as an unusual delay
in getting out of bed—the system notifies the assigned caregiver discreetly, without audible alarms.

This targeted, low-disruption communication reduces unnecessary room visits, allowing staff to prioritize critical care tasks and spend more time on direct
patient interaction. The system leverages scalable salience by initiating visual
updates on staff terminals instead of auditory alarms for non-critical deviations
(e.g., bed-exit delays). Field observations confirmed that caregivers responded

within a median time of 4.5 minutes, demonstrating the effectiveness of low-salience cues in time-sensitive but non-emergency contexts. This finding aligns with Endsley's model of progressive Situation Awareness [6] and supports the hypothesis that calm escalation can maintain responsiveness without inducing alert fatigue. These use cases demonstrate the versatility of Silent Interfaces in balancing proactive support with minimal intrusion, thereby addressing both the functional and emotional dimensions of care technology acceptance.

7   Evaluation and Lessons Learned

The insights in this section are based on practical deployments and iterative feedback loops in real-world assisted living environments, rather than on formal, controlled studies. Data collection combined system log analysis with informal interviews and on-site discussions with residents, caregivers, and facility managers. While the findings are not intended as statistically representative, they provide valuable indications of system behavior and user perceptions under everyday operating conditions.

7.1   Acceptance Indicators

During pilot operations at WoQuaZ, user feedback was collected through semi-structured interviews ($N = 12$ residents, $N = 5$ caregivers, $N = 2$ facility managers). Thematic analysis revealed three dominant indicators of acceptance:

- **Perceived intrusiveness:** Most residents stated that the system did not interfere with their daily routines.
- **Trust in alerts:** Caregivers consistently expressed high trust in the validity of system notifications. They emphasized that the alerts triggered by the system were "reliable" and "rarely false," which increased their confidence in delegating certain monitoring tasks.
- **Overall satisfaction:** Across stakeholder groups, subjective satisfaction ratings were high, particularly in relation to reduced alarm fatigue and minimal disruption of routines.

While these insights are qualitative and based on a limited sample, their consistency across multiple iterations suggests a positive trend in user perception and system acceptance.

7.2   Operational Observations

System log data from uCORE deployments indicated zero false positives in autonomous emergency call scenarios over a cumulative runtime of approximately eight operating months across three installations. Stakeholders confirmed that critical events (e.g., falls, prolonged inactivity) were recognized in a timely manner and appropriately escalated.

In parallel, facility managers observed reductions in energy usage following the deployment of context-sensitive cues (e.g., ventilation and lighting recommendations). Energy monitoring data from the WoQuaZ facility indicate that total annual energy consumption in 2024 was 56% of the baseline recorded in 2012. While multiple factors (e.g., building upgrades, behavioral changes) contributed to this reduction, interviews suggest that Silent Interface cues played a supportive role. These operational insights highlight the real-world viability of multimodal, low-salience feedback mechanisms in both care and sustainability domains, though further controlled studies are needed to isolate causal effects.

7.3   Case Example: The "Washing Machine Load Spike"

A practical lesson emerged during deployment in the WoQuaZ living lab: a reduced number of shared washing machines—introduced to free space for other amenities—led to synchronized use around midday. This created a sharp electricity demand peak, overloading local energy systems and requiring grid power imports.

While the anomaly was detected, the original interaction design did not escalate it due to its categorization as a non-critical event. Because the building's physical layout and appliance placement could not realistically be changed, the solution was a technical workaround: refining event classification so that similar operational anomalies could be escalated when they risk impacting building-wide comfort or safety.

The main takeaway, however, was the importance of applying deeper contextual inquiry during the planning phase. More extensive observation of resident routines before deployment could have revealed the likelihood of synchronized appliance use and allowed preventive design adjustments at the interaction or infrastructure level.

This case illustrates the importance of deep contextual inquiry during early design stages. Had user routines been studied more extensively prior to deployment (e.g., through shadowing or diary methods), this synchrony could have been anticipated and integrated into the escalation logic.

7.4   Key Lessons Learned

Based on field deployments, several key design and operational lessons were identified:

1. **Precision must be paired with prioritization:** Accurate sensor readings alone are insufficient without intelligent, context-aware escalation. This was evident in the "Washing Machine Load Spike" case, where an undetected criticality gap highlighted the need for dynamic reclassification.
2. **Stakeholder diversity matters:** The system's success depended heavily on its ability to support different roles—residents, caregivers, facility managers— via adaptive feedback channels and interaction modalities.

3. **Iteration is essential:** Edge cases and unexpected behavior, such as time-based usage patterns or comfort thresholds, only emerged during in-situ deployment and could not have been predicted through lab-based testing alone.

8   Discussion

The implementation of Silent Interfaces in AAL environments demonstrates that unobtrusive interaction paradigms can achieve high acceptance and operational effectiveness when combined with precise, context-aware event interpretation. However, the broader applicability, long-term viability, and ethical alignment of such systems require careful examination. This section outlines key trade-offs, potential transfer domains, ethical implications, and known limitations, serving as a foundation for future research and system evolution.

8.1   Trade-Offs Between Visibility and Unobtrusiveness

The core strength of Silent Interfaces—their unobtrusive presence and ability to remain in the background—also presents a critical design challenge. Excessive invisibility can lead to reduced user awareness or trust in the system's capabilities, potentially resulting in underutilization. Conversely, increasing visibility risks introducing cognitive load and interaction fatigue, undermining long-term acceptance. Balancing this trade-off requires adaptive interface behavior that dynamically responds to both the urgency of the situation and the preferences of individual users.

Design mechanisms such as progressive disclosure, scalable salience, and context-sensitive feedback loops offer practical strategies for achieving this balance. These approaches align with Endsley's model of progressive situation awareness [6], which emphasizes the importance of timely and contextually appropriate information presentation relative to task demands.

8.2   Scalability to Other Domains

While the current implementation focuses on AAL environments, the underlying design philosophy of Silent Interfaces—context-sensitive, low-interruption interaction—has high potential for cross-domain application. Potential areas include:

- **Healthcare:** Hospital patient monitoring systems that reduce alarm fatigue among medical staff by employing tiered escalation strategies.
- **Industrial safety:** Predictive maintenance systems that only alert operators when conditions require intervention.
- **Smart workplaces:** Energy optimization systems that provide subtle cues for resource-efficient behavior.

– **Community support networks:** Integration with volunteer-platforms such as Hub4Help (currently under development), which connect volunteer helpers with people in need, enabling Silent Interfaces to trigger assistance requests directly to trusted community members.
– **Commercial services:** Connection to paid service providers, such as home maintenance, delivery, or care-on-demand services, allowing automated and context-driven service requests based on detected needs.

In each case, domain-specific rule sets, ontologies, sensing modalities, and interaction patterns would need to be adapted, but the underlying design philosophy remains applicable. By incorporating both community-based volunteer networks and commercial service providers, Silent Interfaces can extend their impact beyond safety and comfort to include proactive, socially integrated support ecosystems. The ability to embed such systems in hybrid ecosystems of public, private, and community services offers a promising pathway for societal impact beyond individual care scenarios.

8.3   Ethical Considerations

The deployment of Silent Interfaces in care environments surfaces several ethical dimensions that must be addressed proactively:

– **Autonomy:** Interventions must remain assistive rather than prescriptive. Users should retain final agency in routine and critical situations.
– **Privacy:** Data collection is minimized by design, leveraging on-device processing wherever feasible. This approach aligns with principles of data minimization and privacy-by-design.
– **Accountability and transparency:** Particularly in autonomous escalation (e.g., emergency calls), it is essential that users and caregivers can understand how decisions are made. Transparent logging and explainable interaction histories support this requirement.

The integration of ethical assessment tools, such as the EVA (Ethical Value Assessment) model, during early design phases has proven effective in translating abstract ethical principles into concrete system requirements and stakeholder dialogues [3].

8.4   Limitations

Despite promising results, current research and system implementation face several limitations.

1. **Geographical and contextual scope:** Pilot deployments were limited to AAL facilities in a single regional context. Broader deployments across different cultural, architectural, and regulatory environments are needed to assess generalizability.

2. **Long-term behavioral adaptation:** The persistence of user responsiveness to subtle cues over extended timeframes remains unverified. Habituation effects could reduce system effectiveness and require recalibration strategies.
3. **Detection scope and classification errors:** While false positives in critical events (e.g., fall detection) were effectively eliminated, operational anomalies—such as irregular utility usage—still present classification challenges. Ongoing refinement of semantic rule sets and hybrid AI models is required.

Addressing these limitations in future work will involve longitudinal studies, cross-domain pilots, and the expansion of participatory design approaches to ensure continued alignment with stakeholder needs and evolving ethical expectations.

9   Conclusion and Future Work

This paper introduced the concept of Silent Interfaces for Active Assisted Living (AAL) environments, combining unobtrusive interaction design with precise, AI-driven situation awareness. Grounded in principles from Calm Technology and Situation Awareness models, and implemented within the uCORE platform, Silent Interfaces demonstrated that it is possible to provide proactive support without overwhelming users with constant interaction.

Through deployments in assisted living facilities, the approach achieved:

- High user acceptance across residents, caregivers, and other stakeholders.
- Certified zero false-alarm rates for autonomous emergency calls.
- Tangible benefits in comfort and energy efficiency through context-triggered cues.

These outcomes confirm that unobtrusiveness and reliability are not mutually exclusive, but can be mutually reinforcing when supported by robust technical architectures and human-centered design processes.

9.1   Future Work

Future research will address several open questions:

1. **Long-term studies:** Investigating how user behavior and trust evolve over extended deployment periods.
2. **Cross-domain applicability:** Adapting Silent Interfaces for healthcare, industrial safety, and workplace environments.
3. **Adaptive personalization:** Refining AI models to continuously learn and adjust engagement thresholds to individual user preferences and contexts.
4. **Expanded ethical evaluation:** Integrating systematic, longitudinal ethical assessments to track the societal and individual impact of Silent Interfaces.

By pursuing these directions, we aim to extend the scalability and versatility of Silent Interfaces, enabling their adoption in a wider range of settings while maintaining the core principles of trust, autonomy, and unobtrusiveness.

Acknowledgments. The authors gratefully acknowledge the foundational work from more than a decade of research in Active Assisted Living (AAL), in particular the Fraunhofer-led AAL projects that have significantly shaped the current approach. Special thanks go to the WoQuaZ living lab in Weiterstadt and the universAAL initiative, whose concepts and findings have had a substantial influence on the technical and design framework presented here.

This work also builds on insights gained through participation in the ongoing, publicly funded AI4ActiveAge project, which explores AI-driven assistance solutions for active ageing. Additional input was provided by the KISCon project, part of the BMBF Incubator initiative, which addresses key challenges in privacy, trust, and security in context-aware ambient systems.

Disclosure of Interests. The authors declare no competing interests relevant to the content of this article.

References

1. Alkhomsan, M.N., Hossain, M.A., Rahman, S.M.M., Masud, M.: Situation awareness in ambient assisted living for smart healthcare. IEEE Access **5**, 20716–20725 (2017). https://doi.org/10.1109/ACCESS.2017.2731363
2. Amershi, S. et aL.: Power to the people: the role of humans in interactive machine learning. AI Mag. **35**(4), 105–120 (2014). https://doi.org/10.1609/aimag.v35i4.2513
3. Bente, S.: Mindsets und methoden für eine praxisnahe digitale ethik. iX – Magazin für professionelle Informationstechnik **1**(8) (2020). https://www.heise.de/select/ix/2020/8/2006210391529546421
4. Cicirelli, G., Marani, R., Petitti, A., Milella, A., D'Orazio, T.: Ambient assisted living: a review of technologies, methodologies and future perspectives for healthy aging of population. Sensors **21**(10), 3549 (2021). https://doi.org/10.3390/s21103549
5. Eggen, B., van den Hoven, E., Terken, J.: Human-centered design and smart homes: how to study and design for the home experience? In: van Hoof, J., Demiris, G., Wouters, E.J.M. (eds.) Handbook of Smart Homes, Health Care and Well-Being, pp. 83–92. Springer, Cham (2017). https://doi.org/10.1007/978-3-319-01583-5_6
6. Endsley, M.R.: Design and Evaluation for Situation Awareness Enhancement, 2nd edn. CRC Press (2016)
7. Ferro, E., et al: The UniversAAL platform for AAL (ambient assisted living). J. Intell. Syst. **24**(3), 301–319 (2015). https://doi.org/10.1515/jisys-2014-0127
8. Hassenzahl, M., Burmester, M., Koller, F.: AttrakDiff: Ein fragebogen zur messung wahrgenommener hedonischer und pragmatischer qualität. Mensch Comput., 187–196 (2003)
9. Hoff, A.: Deployment site AI4ActiveAge – woquaz (2023), aCTIVAGE (BMW IoT Large-Scale Pilot). https://gat.hszg.de/projekte-publikationen/aktuelle-projekte/ai4activeage. Accessed 04 Aug 2025

10. Hoffman, R.R., Mueller, S.T., Klein, G., Litman, J.: Metrics for explainable AI: challenges and prospects. arXiv preprint arXiv:1812.04608 (2018). https://arxiv.org/abs/1812.04608

11. International Organization for Standardization: ISO 9241-210:2019 ergonomics of human-system interaction — human-centred design for interactive systems (2019). https://www.iso.org/standard/77520.html

12. Laugwitz, B., Held, T., Schrepp, M.: Construction and evaluation of a user experience questionnaire. In: Holzinger, A. (ed.) USAB 2008. LNCS, vol. 5298, pp. 63–76. Springer, Heidelberg (2008). https://doi.org/10.1007/978-3-540-89350-9_6

13. Rosson, M.B., Carroll, J.M.: Scenario-based design. In: The Human-Computer Interaction Handbook, pp. 1032–1050. Lawrence Erlbaum Associates (2002)

14. universAAL IoT Coalition: The WoQuaZ experience (2018). Agile Ageing Alliance / universAAL IoT. https://www.agileageing.org/site_files/5944/upload_files/WoQuaZ.pdf. Accessed 04 Aug 2025

15. W3C: Web content accessibility guidelines (WCAG) 2.1 (2018). https://www.w3.org/TR/WCAG21/

16. Weber, K.: MEESTAR: Ein Modell zur ethischen Evaluierung sozio-technischer Arrangements in der Pflege- und Gesundheitsversorgung. In: Technisierung des Alltags. Franz Steiner Verlag (2015)

17. Weiser, M.: The computer for the 21st century. Sci. Am. **265**(3), 94–104 (1991)

18. Weiser, M., Brown, J.S.: The coming age of calm technology. In: Beyond Calculation: The Next Fifty Years of Computing, pp. 75–85. Springer, New York (1997). https://doi.org/10.1007/978-1-4612-0685-9_6

Interactive AI for Preventive Health: Personalization, Gamification, and Ethics (IAI4PH 2025)

Designing an LLM-Powered Social Robot for Supporting Emotion Regulation in Parent-Child Dyads

Jing Li[1(✉)] , Felix Schijve[1] , Sheng Li[2] , Emilia Barakova[1] , and Jun Hu[1]

[1] Department of Industrial Design, Eindhoven University of Technology, Eindhoven 5612, AZ, The Netherlands
`j.li2@tue.nl`, `felixschijve@hotmail.com`, `e.i.barakova@tue.nl`, `J.Hu@tue.nl`
[2] Department of Engineering, Institute of Science Tokyo, Yokohama, Japan
`sheng.li@ieee.org`

Abstract. Emotion regulation (ER) is an essential skill that significantly impacts children's social and emotional development. While prior research has shown that parents play a critical role in shaping their children's ER abilities, the challenges and complexity in parental-involved ER interactions call for technology that delivers context-aware and personalized social support. Although social robotics and Large Language Models (LLMs) both show promise for ER facilitation, few systems integrate language-based reasoning with embodied actions to address mental health needs effectively. To expand the potential applications, we developed an LLM-powered robotic system to facilitate ER in parent-child dyads. We adopt a supervised autonomy approach to integrate natural language dialogues with physical robotic behaviors for multimodal interactions. We detail the technical implementation and interaction design of the system, along with preliminary user tests involving six parent–child dyads. The findings highlight the positive user engagement and trust in interactions with the LLM-powered social robot. Accordingly, we discuss design insights and implications in developing LLM-powered multimodal and autonomous social robot systems for family-centered mental health applications.

Keywords: Social Robot · Large Language Model (LLM) · Emotion Regulation

1 Introduction

Emotion regulation (ER) is a multifaceted concept encompassing emotional awareness, management, and expression [6]. The cultivation of ER skills is vital for a child's social and emotional development [4]. Prior research consistently demonstrates that parents play a central role in shaping their children's ER abilities [2], highlighting a strong positive correlation between the quality of

C. Fayollas et al. (Eds.): EICS 2025, LNCS 16511, pp. 203–211, 2026.
https://doi.org/10.1007/978-3-032-26051-2_15

parent-child interactions and children's emotional competencies [25]. Furthermore, the way parents handle their own emotions has significant implications for how children learn to regulate theirs [5]. When parents find it difficult to identify and articulate their feelings, they may also struggle to acknowledge and respond to their child's emotional needs, particularly in more challenging situations, such as those involving children with psychiatric conditions [13].

Recent research has underscored the value of technological interventions to assist parents in fostering ER in children [23]. These technological supports include mobile applications [24], biofeedback agents [14], and interactive toys [9]. In addition, studies have shown that social support, through positive interactions, affirmation, and emotional assistance, improves both physical and mental outcomes [1]. Social robots provide a unique social presence that can facilitate and enhance parent-child interaction and emotion regulation in various contexts [7]. Indeed, active parental involvement is a necessary condition for effective child ER [19], and multiple evidence-based ER therapies [3] emphasize that parents must also learn to manage their own emotions (e.g., frustration) to project calm, reassuring responses. Against this backdrop, a social robot that interacts with both parents and children can provide holistic ER support [22]. However, the complexity of these interactions demands flexibility and personalized approaches to interact with participants from such diverse user groups as parents and their children [12].

LLMs offer new possibilities for social robots by enabling more context-aware and flexible dialogue [12]. Yet, bridging the gap between LLMs and robotic systems poses significant challenges and requires integrating language-based reasoning with embodied actions. Despite the immense potential of LLM-powered robots to transform human-robot interaction, research is needed to explore design applications and evaluate LLM-powered robotic interactions even in the relatively narrow context of therapeutic applications [12]. In the context of supporting parent-child dyads with ER, these challenges call for an elaborate design approach. We therefore frame the following research questions: RQ1: How can robot-mediated interactions be designed to support ER in parent-child dyads? RQ2: How can LLMs be integrated into a social robotic system for interacting with parent-child dyads?

To address these questions, we propose a MiRo-E social robot system, built around the commercially available MiRo-E platform[1]. In our design, we incorporate evidence-based ER strategies into LLM prompts and robotic behaviors, while using a supervised autonomy approach to support natural language dialogue and multimodal interactions among parent, child, and robot.

The contributions of this paper are twofold. First, we detail a novel integration of LLMs and social robotic technologies to facilitate ER in parent-child dyads. Second, we demonstrate the implementation of evidence-based ER strategies within a supervised autonomous robotic system. The details of the designed system are described in the following sections.

---

[1] https://www.miro-e.com/.

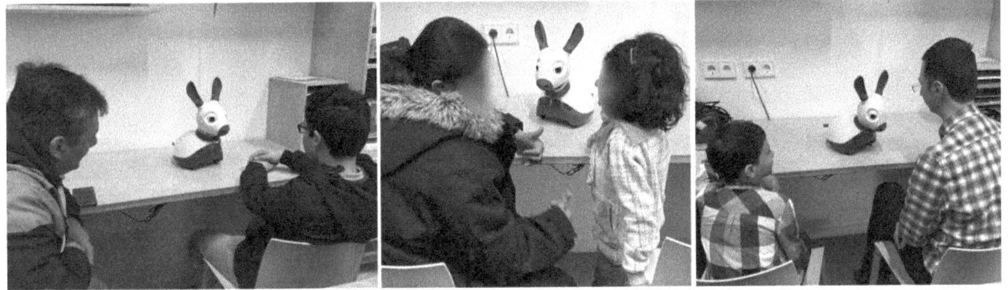

Fig. 1. Parent-child dyads are interacting with MiRo-E.

2   Designing an LLM-Powered Social Robot for Facilitating Parent-Child Dyadic Emotion Regulation

2.1   Designing the Robot-Mediated ER

Evidence-based therapies that focus on ER in the context of parent-child interactions, such as PCIT [3], and Child-Parent Psychotherapy (CPP) [16], share several key strategies. These include: 1) teaching parents to recognize, label and validate their children's lower-intensity emotions while offering coping strategies; 2) guiding parents to regulate their own emotions (for example, frustration) so that they project a calm and reassuring presence; 3) encouraging and praising the use of calm expression or coping techniques by children (for example, breathing exercises); 4) providing psychoeducation on understanding and validating negative emotions; and 5) offering structured practice and live feedback from a therapist.

Drawing on these strategies, our design goal is to incorporate these ER strategies into the autonomous functions of a social robot. Therefore, we developed a six-stage ER procedure in which an LLM-powered social robot actively facilitates parent-child interactions. At the first stage, the robot gets to know the dyad, helping them become comfortable with its presence and interactive features— thereby reducing novelty effects. At the second stage, called *emotion reflection*, the robot prompts the dyad to discuss and reflect on recent negative emotions or stressful events in their everyday lives. The third stage focuses on providing ER education, teaching parents and children how to validate each other's emotional experiences and utilize specific strategies to manage stress. The dyad then practices these techniques under the guidance of a robot during *ER in practice* (the fourth stage). Before the robot ends the conversation in the final stage, it recaps the core ideas of the session, reinforcing the newly learned knowledge at the fifth stage.

2.2   The Proposed Robotics System

To increase parental involvement and support a supervised autonomy approach in social robot therapies, we propose an LLM-based system that interprets semantic content and engages parent-child dyads in context-aware interactions alongside human oversight through robot controllers. Our interactive robotic system comprises several key components: 1) Speech Communication Module with Embedded LLM: This module processes live conversations between the parent-child dyad and the social robot interface. It transcribes and interprets spoken language through the LLM, which generates the robot's verbal feedback. The LLM also extracts the required context from transcribed text, forwarding these as commands to the next component. 2) Behavior Model: Based on commands from the LLM, the Behavior Model activates a set of preprogrammed verbal and non-verbal behaviors. These can be triggered either by Speech Communication Module or by physical cues detected through the embedded sensors of the robot. 3) Robot Controller (Human Operator): For ethical and safety considerations, a human operator supervises the system and acts as a safeguard. This operator can pause or terminate any feedback if needed, ensuring that the LLM will not provide misleading or inappropriate information. The operator also can help maintain momentum between different stages of the therapy, making sure the session progresses smoothly.

Implementation of Speech Communication Module in MiRo-E. The language models enabling MiRo-E's speech and communication capabilities included Whisper[2] for speech recognition, ChatGPT-4o[3] as the LLM, and Google's text-to-speech API[4] for its voice. These were implemented using the OpenAI[5] and Google Cloud[6] API's in Python. The system follows a cascade architecture. First, the microphone captures speech streams, on which voice activity detection (VAD) is performed using WebRTC[7] to isolate the voice input. This ensures that only detected speech is processed by the Whisper model. Whisper then predicts text captions using its decoder, configured to transcribe English text. The transcribed text is subsequently fed into the LLM, which processes the input and generates a textual response. In multi-turn conversations, ChatGPT-4o considers the entire conversation history, allowing it to maintain context and coherence across dialogue exchanges. The generated textual responses are then converted into speech using the Google Cloud text-to-speech system, providing audio feedback to users.

Designing Robotic Behaviors and Expressions. To prompt ChatGPT-4o through each of the six ER stages, we developed prompts that address both the overarching requirements and the specific objectives of every stage. Drawing on established strategies for promoting emotional self-disclosure in social robot

---

[2] https://platform.openai.com/docs/guides/speech-to-text.
[3] https://platform.openai.com/docs/models/gpt-4o.
[4] https://cloud.google.com/text-to-speech?hl=en.
[5] https://openai.com/api/.
[6] https://cloud.google.com/apis?hl=en.
[7] https://webrtc.org/.

therapy, we incorporated *humorous* [11], *vulnerable* [17], and *empathetic* [10] communication styles into our prompt-generation process.

The different emotional expressions and behavioral strategies used by MiRo-E. These context-sensitive behaviors align with the LLM prompts, allowing the LLM to send specific commands (behavioral codes) that trigger certain body movements or vocal expressions. Additionally, a set of preprogrammed rules, described in the following section, can also activate these behavioral feedback. We have defined six behavioral codes that cover various basic emotions and ER strategies: *"Happy/Content"* and *"Upset/Sad"* are commonly used in ER therapies, it helps children recognize positive and negative emotions while fostering empathy and trust [20]. *"Celebration for progress"* is designed with special dancing movements to celebrate milestones or the completion of tasks/ER stages, offering positive reinforcement. The expression *"Confused/ Curious"* indicates MiRo-E's need for more information, which encourages the dyad to self-disclose and interact further. *"Focusing/ Paying attention"* depicts MiRo-E's ongoing engagement with the dyad. This behavior is active by default, unless another expression takes precedence. *"Sleepy"* represents MiRo-E's dormant state, triggered when human controllers pause the communication module or when the ER session ends.

2.3   Interaction Diagram of the Proposed Social Robotic System

To implement a supervised autonomy approach in social robot–facilitated ER sessions, we integrated an LLM-powered speech communication module with a robotic behavior model on the MiRo-E platform, under the supervision of a robot controller. This human controller monitors the interactions between the dyad and MiRo-E while allowing the robot to work toward the defined ER goals.

Figure 2 illustrates the interaction flow. A parent-child dyad engages with MiRo-E through their verbal and non-verbal actions. On the left side of the figure, the LLM-powered speech module converses with the dyad, employing a unique goal-oriented prompt for each stage. The LLM extracts contextual information from the conversation and concludes each stage with a behavioral code. This code then triggers a set of preprogrammed robot behaviors and expressions.

Parents and children also interact physically with MiRo-E in parallel with their verbal input. The behavior model processes gestures such as petting, which activate the *"Happy/Content"* expression. During the interaction, MiRo-E continuously searches for faces during conversations. When MiRo-E detects a face, it switches to a *"Focusing"* expression to show attentive engagement. At the end of the ER session, MiRo-E transitions into a *"Sleep"* mode after bidding the dyad goodbye.

3   User Test

To evaluate the usability of our proposed robotic system, we conducted a laboratory-based user test involving six parent–child dyads (with children aged

6–12). The user test received approval from the Ethics Review Board at Eindhoven University of Technology, Approval Ref. ERB2024ID583.

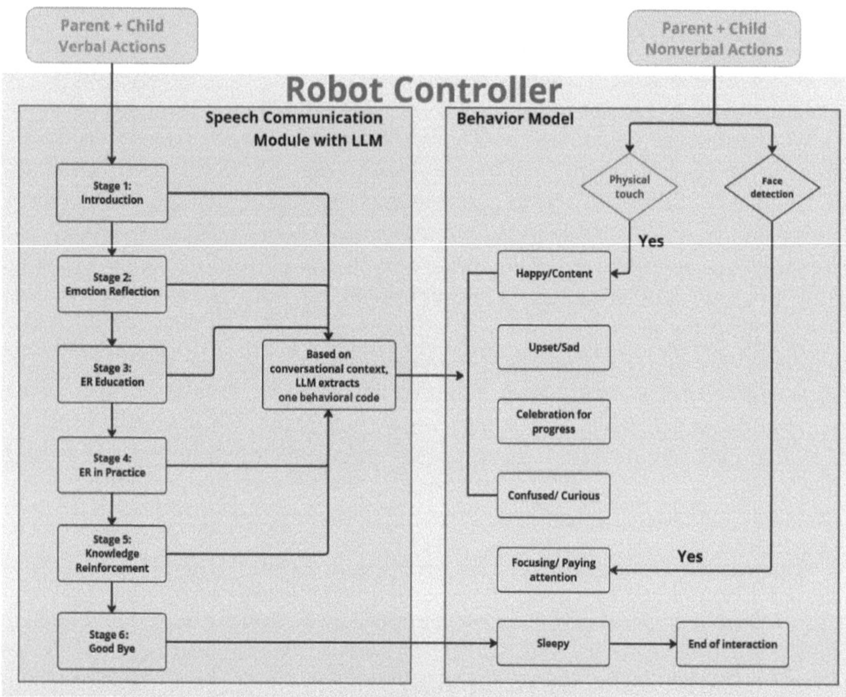

Fig. 2. Interaction diagram of the proposed social robotic system in the ER session

As shown in Fig. 1, during the ER session, MiRo-E initiated the interaction by introducing itself and then worked towards subsequent stages guided by the designed LLM prompts. Although there was no strict time limit, the experimenter could accelerate the process by prompting MiRo-E to move on to the next stage as needed.

We recorded both video and LLM-based conversations throughout the test. After the session, we conducted a semi-structured interview with the parent and child simultaneously, aiming to understand their user experience in terms of user engagement and trust in the robotic system. The interview topics were adapted from the User Engagement Scale [18] and the HRI Trust Perception Scale [21]. The analysis of the interviews and the corresponding video observation identified several preliminary findings.

4   Discussion and Conclusion

Our proposed robotic system employs LLMs as robot policies to deliver ER support and context-aware multimodal feedback, resulting in notably positive human-robot interactions. By enabling natural language dialogue and an extensive repository of ER technique, MiRo-E provided personalized responses tailored to each user's cognitive and emotional state. During user tests, MiRo-E successfully engaged parent-child dyads and fostered trust by demonstrating empathy and incorporating both conversational and physical forms of interaction. Nevertheless, there are several design challenges and implications that warrant further discussion.

We observed different patterns of human-robot interaction. In Fig. 1, for example, the leftmost dyad featured the child speaking directly with MiRo-E while the father mostly observed. In the middle dyad, parent and child both regularly engaged with MiRo-E, creating active triadic interactions. By contrast, the rightmost dyad involved the parent talking to MiRo-E more frequently and facilitating the child's interaction with the robot. These different dynamics in robot-mediated triadic interactions can be influenced by inherent parent-child relationships, participants' personality, and the robot's roles and behaviors. Although we did not explicitly define the role of MiRo-E to the participants, most parents perceived it as a blend of 'psychologist' and 'friend', while children tended to view it simply as a friendly companion.

There are a few technical challenges that may affect how the triadic interaction unfolds. First, MiRo-E currently is unable to distinguish between multiple speakers during a conversation. Although we have designed LLM prompts to specifically target either the parent or the child at different ER stages, MiRo-E sometimes merges inputs when the parent and child speak at the same time, treating them as if they were from a single speaker. Second, the speech communication module is a turn-based system, requiring one party to wait until MiRo-E finishes responding before continuing. As a result, interactions typically take on a dyadic pattern (human-robot or human-human), rather than a fluid three-way exchange among the parent, child, and robot.

Since we aim to support ER for potentially vulnerable individuals, ensuring safety in these interactions is paramount. Previous work has raised concerns about the explainability of LLM [8], data safety and privacy [15], and hallucination behaviors [26] that could inadvertently harm users. To address these ethical concerns and risks, we developed the supervised autonomy approach with human oversight and predefined ER goals, which restricts MiRo-E's autonomy and protects participants' private information. We also prompted LLM to maintain a clear, simple communication style and designed MiRo-E with evidence-based behaviors to enhance clarity and ensure theoretical safety in interactions. In the future, we plan to further iterate on this social robotic system to support parent-child dyads in targeted ER contexts, such as facilitating dyadic interactions during stressful collaborations.

References

1. Akkuş, K., Peker, M.: Exploring the relationship between interpersonal emotion regulation and social anxiety symptoms: the mediating role of negative mood regulation expectancies. Cogn. Ther. Res. **46**(2), 287–301 (2022). https://doi.org/10.1007/s10608-021-10262-0
2. Cooke, J.E., Kochendorfer, L.B., Stuart-Parrigon, K.L., Koehn, A.J., Kerns, K.A.: Parent-child attachment and children's experience and regulation of emotion: A meta-analytic review. Emotion **19**(6), 1103 (2019). https://doi.org/10.1037/emo0000504
3. Eyberg, S.: Parent-child interaction therapy: integration of traditional and behavioral concerns. Child Family Behav. Therapy **10**(1), 33–46 (1988). https://doi.org/10.1300/J019v10n01_04
4. Ferreira, T., Matias, M., Carvalho, H., Matos, P.M.: Parent-partner and parent-child attachment: Links to children's emotion regulation. J. Appl. Dev. Psychol. **91**, 101617 (2024). https://doi.org/10.1016/j.appdev.2023.101617
5. Grabell, A.S., Huppert, T.J., Fishburn, F.A., Li, Y., Hlutkowsky, C.O., Jones, H.M., et al.: Neural correlates of early deliberate emotion regulation: young children's responses to interpersonal scaffolding. Dev. Cogn. Neurosci. **40**, 100708 (2019). https://doi.org/10.1016/j.dcn.2019.100708
6. Gross, J.J.: Emotion regulation: conceptual and theoretical foundations. In: Gross, J.J. (ed.) Handbook of Emotion Regulation, pp. 3–24. Guilford Publications, 2nd edn. (2013)
7. Gvirsman, O., Gordon, G.: Effect of social robot's role and behavior on parent-toddler interaction. In: Proceedings of the 2024 ACM/IEEE International Conference on Human-Robot Interaction, HRI 2024, pp. 222–230. ACM (2024). https://doi.org/10.1145/3610977.3634928
8. Huang, X., Ruan, W., Huang, W., Jin, G., Dong, Y., Wu, C., et al.: A survey of safety and trustworthiness of large language models through the lens of verification and validation. Artif. Intell. Rev. **57**(7), 175 (2024). https://doi.org/10.1007/s10462-024-10824-0
9. Iskanderani, A.: Research Through Design for Anger Co-Emotion Regulation Learning in Young Children: A Soft Toy Design for Deep Breathing. Ph.D. thesis, Te Herenga Waka-Victoria University of Wellington (2023), https://doi.org/10.26686/wgtn.24235927
10. Johanson, D.L., Ahn, H.S., Broadbent, E.: Improving interactions with healthcare robots: a review of communication behaviours in social and healthcare contexts. Int. J. Soc. Robot. **13**(8), 1835–1850 (2021). https://doi.org/10.1007/s12369-020-00719-9
11. Johanson, D.L., Ahn, H.S., Lim, J., Lee, C., Sebaratnam, G., MacDonald, B.A., et al.: Use of humor by a healthcare robot positively affects user perceptions and behavior (2020). https://doi.org/10.1037/tmb0000021
12. Kim, C.Y., Lee, C.P., Mutlu, B.: Understanding large-language model (llm)-powered human-robot interaction. In: Proceedings of the 2024 ACM/IEEE International Conference on Human-Robot Interaction, HRI 2024, pp. 371–380. ACM (2024). https://doi.org/10.1145/3610977.3634966
13. Lench, H.C., Levine, L.J., Whalen, C.K.: Exasperating or exceptional? parents' interpretations of their child's adhd behavior. J. Atten. Disord. **17**(2), 141–151 (2013). https://doi.org/10.1177/1087054711427401

14. Li, J., Wang, P., Barakova, E.I., Hu, J., Dai, G.: Stress diffuser: a biofeedback agent for stress management in children during homework with parent involvement. In: Proceedings of the 23rd Annual ACM Interaction Design and Children Conference, IDC 2024, pp. 701–705. ACM (2024). https://doi.org/10.1145/3628516.3659378
15. Liaw, S.T., Liyanage, H., Kuziemsky, C., Terry, A.L., Schreiber, R., Jonnagaddala, J., et al.: Ethical use of electronic health record data and artificial intelligence: recommendations of the primary care informatics working group of the international medical informatics association. Yearb. Med. Inform. **29**(01), 051–057 (2020). https://doi.org/10.1055/s-0040-1701980
16. Lieberman, A.F., Van Horn, P., Ippen, C.G.: Toward evidence-based treatment: Child-parent psychotherapy with preschoolers exposed to marital violence. J. Am. Acad. Child & Adolescent Psychiatry **44**(12), 1241–1248 (2005). https://www.sciencedirect.com/science/article/pii/S0890856709622358
17. Martelaro, N., Nneji, V.C., Ju, W., Hinds, P.: Tell me more designing hri to encourage more trust, disclosure, and companionship. In: 2016 11th ACM/IEEE International Conference on Human-Robot Interaction (HRI), pp. 181–188. IEEE (2016). https://doi.org/10.1109/HRI.2016.7451750
18. O'Brien, H.L., Cairns, P., Hall, M.: A practical approach to measuring user engagement with the refined user engagement scale (ues) and new ues short form. Int. J. Hum Comput Stud. **112**, 28–39 (2018). https://doi.org/10.1016/j.ijhcs.2018.01.004
19. Piccolo, A., De Domenico, C., Di Cara, M., Settimo, C., Corallo, F., Leonardi, S., et al.: Parental involvement in robot-mediated intervention: a systematic review. Front. Psychol. **15**, 1355901 (2024). https://doi.org/10.3389/fpsyg.2024.1355901
20. Rocha, M., Valentim, P., Barreto, F., Mitjans, A., Cruz-Sandoval, D., Favela, J., et al.: Towards enhancing the multimodal interaction of a social robot to assist children with autism in emotion regulation. In: International Conference on Pervasive Computing Technologies for Healthcare, pp. 398–415. Springer (2021). https://doi.org/10.1007/978-3-030-99194-4_25
21. Schaefer, K.E.: Measuring trust in human robot interactions: Development of the "trust perception scale-hri". In: Robust intelligence and trust in autonomous systems, pp. 191–218. Springer (2016).https://doi.org/10.1007/978-1-4899-7668-0_10
22. Shin, J.Y., Rheu, M., Huh-Yoo, J., Peng, W.: Designing technologies to support parent-child relationships: a review of current findings and suggestions for future directions. Proceedings of the ACM on Human-Computer Interaction, vol.5(CSCW2), pp. 1–31 (2021). https://doi.org/10.1145/3479585
23. Silva, L.M., Cibrian, F.L., Bonang, C., Bhattacharya, A., Min, A., Monteiro, E.M., et al.: Co-designing situated displays for family co-regulation with adhd children. In: Proceedings of the CHI Conference on Human Factors in Computing Systems, pp. 1–19 (2024). https://doi.org/10.1145/3613904.3642745
24. Silva, L.M., Cibrian, F.L., Monteiro, E., Bhattacharya, A., Beltran, J.A., Bonang, C., et al.: Unpacking the lived experiences of smartwatch mediated self and co-regulation with adhd children. In: Proceedings of the 2023 CHI Conference on Human Factors in Computing Systems, pp. 1–19 (2023)
25. Ting, V., Weiss, J.A.: Emotion regulation and parent co-regulation in children with autism spectrum disorder. J. Autism Dev. Disord. **47**, 680–689 (2017). https://doi.org/10.1007/s10803-016-3009-9
26. Zhang, Y., Li, Y., Cui, L., Cai, D., Liu, L., Fu, T., et al.: Siren's song in the ai ocean: a survey on hallucination in large language models. arXiv preprint arXiv:2309.01219 (2023)

Galactic Glide: A Gamified Physical Therapy Rehabilitation System for Adolescent Populations

Ashley Pham[1], Antonio Gonzalez[1], Gabriel Mozombite Gallegos[1],
Julia Woodward[2], Nathan Schilaty[1,3,4],
and John Michael Templeton[1,2,4(✉)]

[1] College of Engineering - Medical Engineering, University of South Florida, Tampa, FL 33620, USA
{ashleypham,agonzalez46,gabrielmozombite,nschilaty,jtemplet}@usf.edu
[2] Bellini College of Artificial Intelligence, Cybersecurity, and Computing, University of South Florida, Tampa, FL 33620, USA
juliaevewoodward@usf.edu
[3] Morsani College of Medicine - Department of Neurosurgery, Brain, and Spine, University of South Florida, Tampa, FL 33613, USA
[4] Center for Neuromusculosketal Research, Morsani College of Medicine, Tampa, FL 33620, USA

Abstract. Artificial Intelligence (AI) and digital devices are becoming more pervasive in the monitoring of individuals' health as device functionalities increase and their presence in daily life continues to grow. Subsequently, these tools can provide transformative healthcare practice by enabling more personalized, proactive, and engaging interventions. This work presents the design and development of an interactive physical therapy system for adolescents that integrates a wearable EMG leg sensor array with a computer-based video game. The system captures EMG signals during therapeutic exercises and uses them to drive in-game mechanics, providing real-time feedback while also aiming to improve therapeutic engagement. A key feature of the system is its ability to adapt based on individual performance and clinician input, allowing personalized therapy progression and reporting. Using this approach, the system supports continuous personalized assessment and customized intervention for children while promoting motivation through dynamic gameplay. The developed system illustrates a promising direction for designing interactive, AI-augmented health technologies that can enhance pediatric rehabilitation outcomes through accessible, game-based experiences. Further, this work contributes to ongoing research in human-computer interaction, AI, and behavioral health by exploring multisensory interactions, social AI elements, and methods for evaluating user engagement.

Keywords: Wearables · Digital Health · Human Computer Interaction · Gamification · Artificial Intelligence

C. Fayollas et al. (Eds.): EICS 2025, LNCS 16511, pp. 212–220, 2026.
https://doi.org/10.1007/978-3-032-26051-2_16

1   Introduction

Pediatric physical disability, particularly those affecting mobility, is a significant public health concern, with thousands of children experiencing impairment due to neuromuscular or traumatic causes [1,2]. According to recent epidemiological data, lower limb conditions (e.g., fractures, cerebral palsy, muscular dystrophies) are among the most common causes of physical disability in adolescents. These conditions frequently result in reduced muscle strength, limited joint mobility, and impaired balance/coordination, often requiring long-term rehabilitation to restore function [3]. Sports-related lower extremity injuries account for a large proportion of cases, with a high prevalence of ankle and knee injuries [4]. For those with developmental or neuromuscular disorders, mobility limitations often persist into adulthood, further underscoring the need for early and effective intervention strategies [5].

Subsequently, engagement in physical rehabilitation is a key determinant of success in restoring or maintaining functional mobility, particularly in pediatric populations facing challenges related to lower limb strength and coordination [6]. Tasks such as ambulating, rising from a seated position, and navigating stairs are foundational to independent living, yet they often present significant hurdles for children undergoing rehabilitation due to neuromuscular or orthopedic conditions [3]. Traditional therapy approaches, while clinically effective, may fall short in maintaining sustained participation, especially in younger patients who benefit from more interactive and motivating therapeutic environments [7].

To address this challenge, we propose the development of a gamified rehabilitation system—*Galactic Glide*—designed to enhance participation and improve outcomes for lower leg mobility exercises in adolescent populations. The system integrates a wearable sensor array capable of capturing electromyography (EMG) signals with an engaging computer-based video game. Through real-time signal collection and feedback, children are able to visualize movements as game actions, reinforcing physical therapy in a way that is both intuitive and enjoyable. Additionally, the system incorporates adaptive personalization that allows for clinician-guided tuning and progression tracking. By aligning therapeutic goals with engaging gamification, the platform aims to sustain long-term motivation and improve functional outcomes of physical therapy for adolescent populations.

2   Related Work

Gamification is the use of game design elements alongside mechanics in non-game contexts to increase engagement, motivation, fun, and passion across tasks [8]. Adding gamification elements (e.g. score, difficulty levels, animations) has been shown to help increase children's motivation and engagement in multiple contexts. For instance, gamification components increase study completion rates for children [9], as well as enhance their learning and engagement for educational applications [10]. Woodward et al. [11] found that gamified elements in children's touchscreen applications can lead to more precise touch interactions due to game immersion. Prior work has also found that children expect intelligent user interfaces (IUIs) to be customized and tailored to user preferences, and that IUIs should include visual screen elements, including animations and effects [12].

Gamification is also being applied in healthcare for children in settings across chronic disease management in addition to rehabilitation, physical activity, and mental health [13–17]. Gamification has been shown to be more entertaining and simple to use in health contexts, especially for children [13]. For instance, Maderia et al. [14] created a mobile game to support speech therapy of phonological development for children ages 3 to 8 years old. Garde et al. [17] created a mobile game for children ages 8 to 13 called *MobileKids Monster Manor* to promote children's physical activity. Pimentel-Ponce et al. [18] conducted a systematic review on the use of gamification for children's neurological motor rehabilitation. They found that gamification could aid in improving children's motivation, balance, strength, function, coordination, and satisfaction. However, a majority of the studies used Nintendo Wii commercial applications, which do not incorporate EMG signals or allow for personalized therapy progression. Overall, gamification has been shown to aid in children's engagement and motivation in healthcare, but their are still open research areas on how to create more personalized and adaptable gamification applications to aid in children's lower leg mobility.

3 System Design

3.1 Wearable Design

The wearable was designed to be lightweight, form-fitting, secure, and compatible with EMG-based signal acquisition. Figure 1(a) depicts all components including electronics (Parts B-D & F) and housings (Parts A, E, & G).

This system consists of a custom-fabricated wearable double-layered neoprene sleeve (Fig. 1(a) - Part E) with a fitted compartment to securely house the MyoWare 2.0 Wireless Shield and Muscle Sensor (Fig. 1(a) - Parts B & C). The sleeve is lined with unbroken loop fabric for adjustable Velcro attachment. Further, it is configured for use on either leg where upward facing right/left arrows allow for corresponding limb use. This layered design was formulated to ensure comfort, modularity, and stable sensor positioning throughout gameplay.

For electronics, the system makes use of several dev-kit components. Primarily, this system uses a MyoWare 2.0 Muscle Sensor (Fig. 1(a) - Part B) which records superficial EMG activity from the user's lower leg via three disposable gel electrodes (Fig. 1(a) - Part D), two for signal detection and one for reference. These surface electrodes adhere to the skin in a standard bipolar electrode configuration on the calf [19] (Fig. 1(c)). The sensor snaps directly into the MyoWare 2.0 Wireless Shield (Fig. 1(a) - Part C), which transmits EMG data wirelessly to an ESP32-WROOM microcontroller (Fig. 1(a) - Part F) that is serially connected to a PC running *Galactic Glide* (Fig. 1(b)). Custom housings were made for consumer dev-kit deployment (Fig. 1(a) - Parts A & G). These custom housings were made to accommodate electronics while maintaining sensor alignment, in addition to embedding clips to prevent device slippage during movement.

The modularity of the design allows easy access for electronic removal and recharging between sessions. The snug fit ensures good electrode contact and minimizes motion artifact, which is critical for clean signal acquisition. In general, this design enables the safe and effective acquisition of EMG signals in a physically secure form factor that supports dynamic use during gameplay.

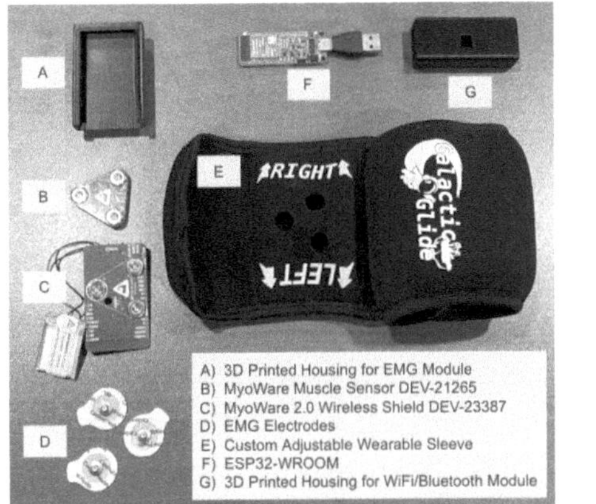

((a)) Wearable hardware components.

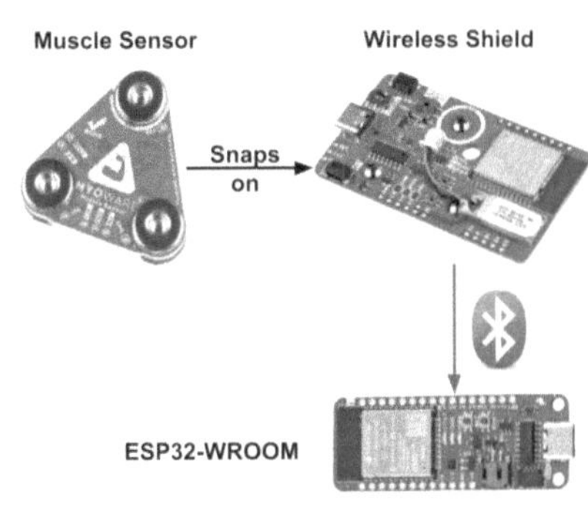

((b)) MyoWare hardware connectivity.

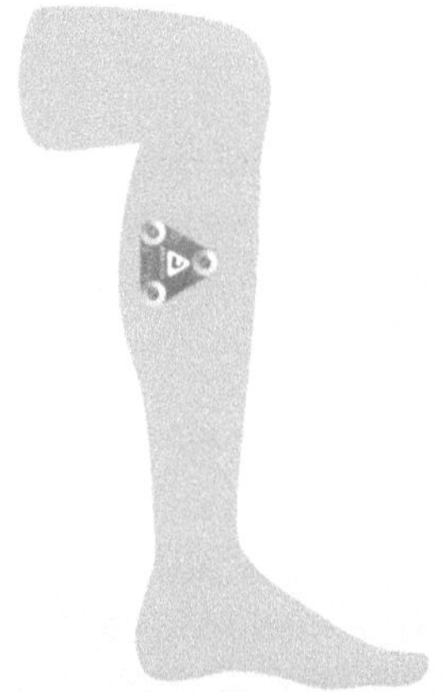

((c)) EMG placement.

Fig. 1. Overview of wearable system hardware and connections.

3.2   Application Design

Data Processing. *Galactic Glide* is developed in Python which integrates live EMG data to control gameplay via PyGame (Fig. 2). To convert raw EMG signals into input, a two-stage signal processing pipeline is used. First, root mean square (RMS) (Eq. 1) is applied to a rolling window of raw EMG values.

$$RMS = \sqrt{\frac{1}{n} \sum_i x_i^2} \tag{1}$$

This reduces noise and provides a stable measure of muscle activation over time. Subsequently, further smoothing is done using a moving average filter to reduce jitter caused by background EMG activity. The result is a continuous signal that reflects user efforts while minimizing artifact and variability.

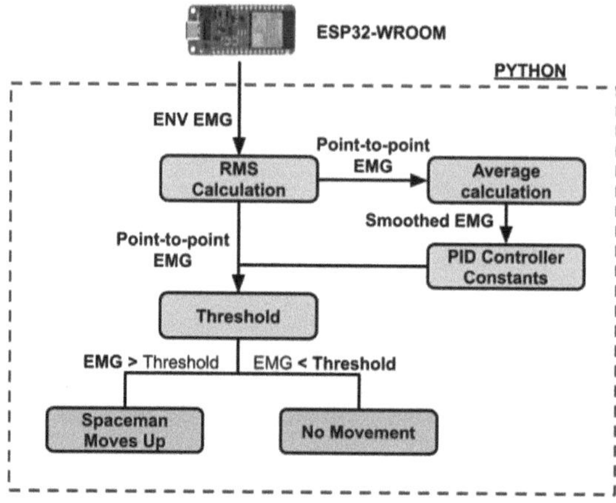

Fig. 2. EMG signal processing pipeline for character activation.

To translate this smoothed signal into responsive in-game motion, the system uses a proportional-integral-derivative (PID) controller (Eq. 2). This continuously adjusts output to help regulate the astronaut's movement in response to EMG activity (e.g., strong/fast muscle activation produces faster movements, with reduced overshooting/instability).

$$u(t) = K_p e(t) + K_i \int_0^t e(\tau)\, d\tau + K_d \frac{de(t)}{dt} \tag{2}$$

Gameplay Mechanics and Engagement. Space was chosen as the context for the game because it is an engaging theme for children. According to a 2019 survey conducted in the US, UK and China, 86% of children aged 8 to 12 are interested in space exploration [20]. Thus, users start by completing calibration (e.g., 'Astronaut Training') to personalize character response (Fig. 3). This calibration collects data regarding maximum EMG activation (e.g., via 'Astro Boost') and control (e.g., via 'Glide Practice') for use in smoothing and personalized thresholds, ensuring fair and consistent gameplay.

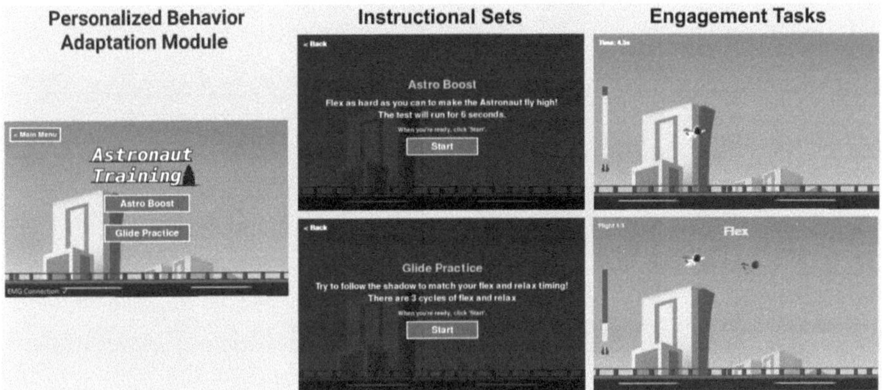

Fig. 3. Depiction of personalized behavior adaptation module components.

Once calibrated, players engage with the interactive gamification module, which includes themed levels of increasing difficulty (Fig. 4). During gameplay, users control an astronaut navigating space-themed environments. A vertical "jet boost" bar is shown as part of interactive gameplay which offers immediate visual feedback of EMG activation. The horizontal bar in the "jet boost" pane shows the threshold to cause character ascension. These features encourage user engagement, while also building a sense of understanding during play.

The game includes five progressively challenging levels (e.g., shown in 'Planet Select' and 'Level Themes' sections of Fig. 4), each represented by a different planet. These levels are progressively unlocked (i.e., users must complete one to access the next) to instill a clear sense of progression and motivate continued engagement. As players advance, they are tasked with increasing EMG activation according to calibrated thresholds. To maintain motivation and reduce discouragement, the game features a multi-life system. Lives reset after the completion of each level allowing users ample opportunity to complete protocols. Further, distinct level themes, sound cues, and responsive feedback make each session engaging, helping players stay consistent with their therapy routines.

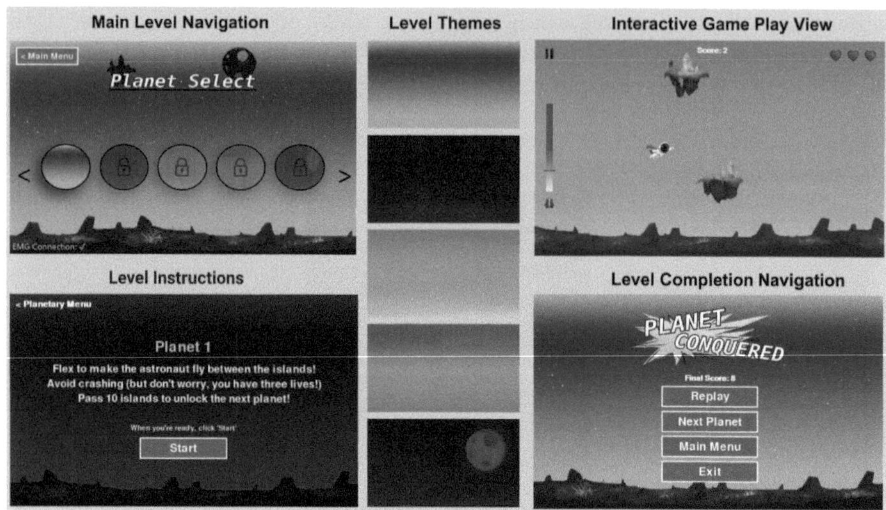

Fig. 4. Depiction of interactive gamification module components.

4   Discussion

4.1   Clinical Relevance

Galactic-Glide is designed with clinical integration as a core goal, addressing key factors for long-term rehabilitation success in adolescents. First, it promotes sustained engagement in repetitive therapeutic exercises by embedding them in an interactive, progressively challenging game, helping maintain motivation—often a barrier in traditional therapy [17]. Second, it personalizes gameplay based on user capability through adaptive calibration (Fig. 3), aligning theraputic outcomes with user-specific EMG activation to ensure appropriate challenge and prevent discouragement. Third, it includes clinician-configurable thresholds, allowing adjustable muscle activation targets, repetition goals, and session durations tailored for individuals. This flexibility supports personalized therapy progression. Fourth, robust data logging captures user engagement, progress, and EMG trends, offering clinicians actionable insights for evaluating adherence and outcomes. These features also enhance interdisciplinary communication and support both home and clinic-based accountability. Altogether, *Galactic-Glide* offers a scalable, adaptable solution for improving the accessibility, consistency, and effectiveness of pediatric lower-limb rehabilitation in real-world settings.

4.2   Limitations and Future Work

While the design of *Galactic Glide* includes key features to enhance rehabilitation engagement and clinical outcomes in adolescents, further opportunities remain. A primary limitation is the lack of comprehensive user testing across diverse adolescent groups, particularly those with varying cognitive abilities and impairment severity. Future work will address this through broad usability and acceptability studies, clinical impact evaluations, and integration with existing workflows. Incorporating clinician feedback, along with longitudinal outcome tracking, will be essential for refining system effectiveness. Continued efforts will also focus on improving accessibility—such as device compatibility and customizable feedback—for at-home use and novel activation sites (e.g., upper limbs), broadening the system's reach across rehabilitation needs. Finally, future work will explore wholistic integration of predictive AI to further personalize therapy.

5   Conclusion

The development of *Galactic Glide* demonstrates the potential of combining wearable biosensing with gamified interfaces to improve pediatric rehabilitation. By translating lower limb EMG signals into meaningful, interactive gameplay, the system not only facilitates engagement in repetitive therapeutic exercises but also introduces a scalable framework for adaptive, clinician-informed therapy. This approach supports both real-time performance feedback and long-term monitoring, aligning with current trends in personalized digital health. As technology continues to reshape the delivery of care, systems like *Galactic Glide* offer a compelling model for empowering young patients to take an active role in recovery, promoting sustained motivation and improved clinical outcomes.

References

1. Liu, H.: Epidemiological evaluation of traumatic lower limb fractures in children: Variation with age, gender, time, and etiology. Medicine **98**(38), e17123 (2019). https://doi.org/10.1097/MD.0000000000017123
2. Theadom, A., Rodrigues, M., Poke, G., O'Grady, G., Love, D., Hammond-Tooke, G., et al.: A nationwide, population-based prevalence study of genetic muscle disorders. Neuroepidemiology **52**(3–4), 128–135 (2019). https://doi.org/10.1159/000494115
3. Goyal, N., Goudriaan, M., Scheys, L., Desloovere, K., Molenaers, G., Huenaerts, C., et al.: Children with bilateral cerebral palsy rely more on hip moments to perform step-up tasks: Implications for rehabilitation. Front. Hum. Neurosci. **18**, 1343457 (2024). https://doi.org/10.3389/fnhum.2024.1343457
4. Emery, C.A., Pasanen, K.: Current trends in sport injury prevention. Best Pract. Res. Clinical Rheumatol. **31**(2), 249–265 (2017). https://doi.org/10.1016/j.berh.2017.03.003
5. Sienko, K.H., Whitney, S.L., Carender, W., Hegeman, J.: Mobility and balance in pediatric populations with sensory deficits: implications for rehabilitation. Phys. Ther. **99**(5), 629–639 (2019). https://doi.org/10.1093/ptj/pzz020
6. Willingham, T.B., Stowell, J., Collier, G., Backus, D.: Leveraging emerging technologies to expand accessibility and improve precision in rehabilitation and exercise for people with disabilities. Int. J. Environ. Res. Public Health **21**(1), 79 (2024). https://doi.org/10.3390/ijerph21010079

7. Ali, S.G., Wang, X., Li, P., Jung, Y., Bi, L., et al.: A systematic review: virtual-reality-based techniques for human exercises and health improvement. Front. Public Health **11**, 1143947 (2023). https://doi.org/10.3389/fpubh.2023.1143947

8. Wood, L.C., Reiners, T.: Gamification. In: Khosrow-Pour, M (ed.) Encyclopedia of Information Science and Technology, Third Edition, pp. 2878–2886. IGI Global (2015). https://doi.org/10.4018/978-1-4666-5888-2.ch297

9. Brewer, R., Anthony, L., Brown, Q., Irwin, G., Nias, J., Tate, B.: Using gamification to motivate children to complete empirical studies in lab environments. In: Proceedings of the 12th International Conference on Interaction Design and Children, pp. 388–391 (2013). https://doi.org/10.1145/2485760.2485816

10. Nand, K., Baghaei, N., Casey, J., Barmada, B., Mehdipour, F., Liang, H.-N.: Engaging children with educational content via gamification. Smart Learn. Environ. **6**, 1–15 (2019). https://doi.org/10.1186/s40561-019-0085-2

11. Woodward, J., et al. Characterizing how interface complexity affects children's touchscreen interactions. In: Proceedings of the 2016 CHI Conference on Human Factors in Computing Systems, pp. 1921–1933 (2016). https://doi.org/10.1145/2858036.2858200

12. Woodward, J., McFadden, Z., Shiver, N., Ben-Hayon, A., Yip, J.C., Anthony, L.: Using co-design to examine how children conceptualize intelligent interfaces. In: Proceedings of the 2018 CHI Conference on Human Factors in Computing Systems, pp. 1–14 (2018). https://doi.org/10.1145/3173574.3174149.

13. Gkintoni, E., Vantaraki, F., Skoulidi, C., Anastassopoulos, P., Vantarakis, A.: Promoting physical and mental health among children and adolescents via gamification–a conceptual systematic review. Behav. Sci. **14**(2), 102 (2024). https://doi.org/10.3390/bs14020102

14. Madeira, R.N., Macedo, P., Reis, S., Ferreira, J.: Super-fon: mobile entertainment to combat phonological disorders in children. In: Proceedings of the 11th Conference on Advances in Computer Entertainment Technology, pp. 1–4 (2014). https://doi.org/10.13140/2.1.1704.8647

15. Sardi, L., Idri, A., Fernández-Alemán, J.L.: A systematic review of gamification in e-health. J. Biomed. Inform. **71**, 31–48 (2017). https://doi.org/10.1016/j.jbi.2017.05.011

16. Keung, C., Lee, A., Lu, S., O'Keefe, M.: Bunnybolt: a mobile fitness app for youth. In: Proceedings of the 12th International Conference on Interaction Design and Children ser. IDC 2013, pp. 585–588. Association for Computing Machinery, New York (2013). https://doi.org/10.1145/2485760.2485871

17. Garde, A., Umedaly, A., Abulnaga, S.M., Robertson, L., Junker, A., Chanoine, J.P., et al.: Assessment of a mobile game ("mobilekids monster manor") to promote physical activity among children. Games Health J. **4**(2), 149–158 (2015). https://doi.org/10.1089/g4h.2014.0095

18. Pimentel-Ponce, M., Romero-Galisteo, R., Palomo-Carrión, R., Pinero-Pinto, E., Merchán-Baeza, J.A., Ruiz-Muñoz, M., et al.: Gamification and neurological motor rehabilitation in children and adolescents: a systematic review. Neurologia (Engl Ed) **39**(1), 63–83 (2024). https://doi.org/10.1016/j.nrleng.2023.12.006

19. Analia, R., Sulistyo, V.A., Iglesias, T.M., Pamungkas, D.S., Jamzuri, E.R.: et al. The reusable electrode of emg sensor for capturing the calf muscle activities. In: 2020 3rd International Conference on Applied Engineering (ICAE), pp. 1–4. IEEE (2020). https://doi.org/10.1109/ICAE50557.2020.9350558

20. Group, T.L. Lego group kicks off global program to inspire the next generation of space explorers as nasa celebrates 50 years of moon landing (Jul 2019)

Avatar-Based Conversational AI for Personalized Physical Rehabilitation in Diabetes and Osteoarthritis Management

Aron Samuel Georgekutty[1]([envelope]) [ORCID], Aarnout Brombacher[2] [ORCID], and Jun Hu[1] [ORCID]

[1] Department of Industrial Design, Eindhoven University of Technology, Eindhoven, The Netherlands
{a.s.georgekutty,j.hu}@tue.nl
[2] Jheronimus Academy of Data Science, 's-Hertogenbosch, The Netherlands
a.c.brombacher@tue.nl

Abstract. The growing global burden of chronic conditions like diabetes and osteoarthritis necessitates innovative, scalable rehabilitation strategies. This paper proposes a conversational AI system designed to deliver dynamic, personalized physical rehabilitation plans. The system leverages large language models (LLMs) and physiotherapist supervision to adapt recommendations based on user feedback and physiological data. A key novelty of this work is its avatar-based AI agent, designed with the potential for integration into extended reality (XR) environments, thus aligning with emerging AR glass technologies. We present an early-stage prototype supported by a modular architecture and propose a pathway toward implementation and evaluation to enhance patient adherence, safety, and engagement in long-term rehabilitation.

Keywords: Conversational AI · Physical Rehabilitation · Diabetes · Osteoarthritis · Personalized Healthcare · XR · LLMs · Avatar

1   Introduction

Chronic conditions such as diabetes mellitus and osteoarthritis are among the leading causes of disability and healthcare costs globally. Diabetes affects over 537 million adults worldwide [1], and osteoarthritis remains the most common musculoskeletal disorder in the elderly. Traditional rehabilitation programs often struggle with limited personalization, insufficient monitoring, and poor long-term adherence.

Although digital health solutions are growing, most lack the adaptability and personalization needed for complex, evolving rehabilitation needs. Our work addresses these gaps by proposing a conversational AI system capable of dynamically generating and adjusting physical activity recommendations based on real-time user interaction and physiotherapist guidance. A unique feature of this

C. Fayollas et al. (Eds.): EICS 2025, LNCS 16511, pp. 221–227, 2026.
https://doi.org/10.1007/978-3-032-26051-2_17

system is its avatar-based interface, which enables more intuitive, immersive interactions and positions the system for future deployment within AR glasses or XR platforms. This paper contributes a conceptual and technical foundation for such a system. Unlike existing AI healthcare solutions limited to symptom-checking or static advice, our system emphasizes real-time dialogue, adaptive feedback loops, professional oversight, and futureproofing via XR integration. We aim to provide a safer, more engaging, and patient-centric rehabilitation experience.

2   Related Work

The application of conversational agents in healthcare has expanded across various domains, including symptom assessment, chronic disease management, and mental health. Foundational examples include Woebot [5], a chatbot for delivering cognitive behavioral therapy, and Replika, a virtual companion focused on emotional support. While these systems illustrate the potential of natural language interaction, they are not tailored for physical rehabilitation tasks.

Conversational agents used for physical rehabilitation are still relatively nascent but rapidly growing. Several early systems have explored limited functionality. For instance, the work by M.H. Lee (2019) presents a rule-based dialogue system for post-stroke rehabilitation, focused on improving adherence through an intelligent decision support system [8]. Similarly, Griffin et al. (2021) evaluated a virtual coach for patients with chronic pain, integrating goal-setting and motivational interviewing techniques [7].

More recent approaches are incorporating multimodal inputs. Boltaboyeva et al. (2022) illustrated how AI, IoT, and LLM-based virtual assistants hold significant promise for addressing current healthcare challenges through their ability to enhance, personalize, and streamline patient care. [3]. Devane et al. (2022) reviewed the use of virtual reality for rehabilitation, emphasizing the need for personalization and empathic interaction [4].

Rainey et al. (2023) demonstrated that a multilingual SMS-based chatbot could effectively engage patients after total joint arthroplasty, particularly those with limited English proficiency, leading to reduced readmissions and improved equity in care delivery [10]. However, these systems often rely on simplified dialogue capabilities and static routines.

Avatar-based conversational agents are also gaining attention. Belpaeme et al. (2018) reviewed social robots and virtual agents in care settings, highlighting their potential to increase engagement, especially when visual embodiment is included [2]. Mercado et al. (2023) explored embodied conversational agents for home-based exercise interventions with encouraging results in usability and adherence [9].

In addition, the integration of XR and AR with AI in rehabilitation is becoming an emerging research direction. Vinolo et al. (2022) systematically reviewed AR-based physical therapy methods and analyzed their effectiveness, and Fregna et al. (2021) demonstrated the benefit of immersive agents in VR for balance training in older adults [11] [6]. These findings suggest that adding spatial and visual context to conversational interaction may improve rehabilitation outcomes.

Despite these advancements, existing systems rarely integrate the full loop: contextual memory from large language models, multimodal sensor fusion, clinical supervision, and immersive visual feedback. Our work addresses this by combining a retrieval-augmented LLM with an avatar-based interface and real-time physiological data to deliver adaptive rehabilitation coaching, with extensibility toward XR environments.

3   System Architecture

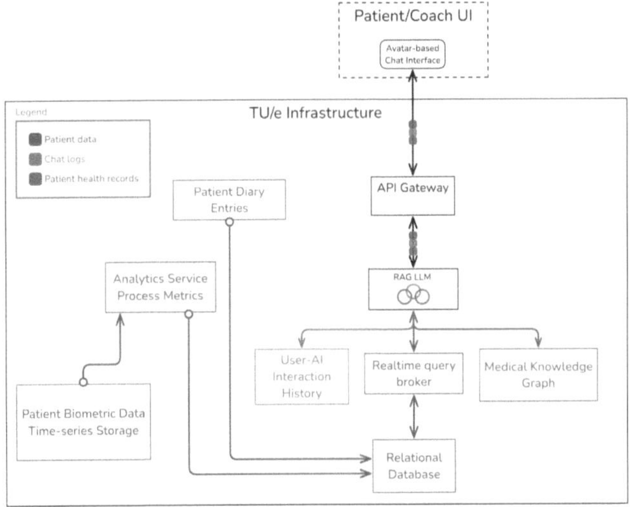

Fig. 1. System architecture for the avatar-based conversational rehabilitation platform.

Figure 1 above illustrates the architecture of the avatar-based conversational rehabilitation platform, designed to enable adaptive, patient-centered care using large language models (LLMs) and multimodal data inputs. The system operates across multiple coordinated layers:

Patient Interface: Patients and coaches interact with an avatar-based interface capable of voice and text communication. This interface is implemented using intuitive UI/UX principles and is designed for future deployment in augmented or extended reality (AR/XR) environments such as AR glasses. All interactions are channeled through a central interaction middleware.

Interaction Middleware (API Gateway): The API Gateway acts as a mediator between the user-facing interface and backend services. It handles data formatting, authentication, and routing between components, ensuring secure and efficient data exchange.

Retrieval-Augmented Generation (LLM Module): At the core of the system is a large language model enhanced with retrieval capabilities. This LLM interprets patient inputs, queries relevant context (from chat logs, patient records, and diaries), and generates context-aware, medically appropriate responses. Prompting strategies and context windows are used to maintain coherence and personalize feedback over time.

Data Sources and Storage: The system draws on several structured and unstructured data sources. These include knowledge graphs from Electronic Medical Records(EMRs) for historical clinical data, user-AI interaction logs to maintain conversational continuity, self-reported diary entries for qualitative insights, and real-time biometric data streams (e.g., heart rate, steps) from wearable devices. These data sources are processed and stored in a central relational database. A real-time cache and messaging system (e.g., Redis) enables fast data retrieval and communication across components.

Analytics Service: To complement the conversational system, an analytics engine processes biometric and behavioral metrics. This component evaluates trends such as user adherence, fatigue, and engagement levels, and can feed back into both the LLM module and clinician oversight workflows. These insights can also support the continuous improvement of the AI recommendations and inform clinical decision-making.

Together, this architecture ensures the system is responsive, adaptive, and scalable. It facilitates a closed-loop rehabilitation framework where patient and coach input and the physiological data directly influence the nature of care delivered. The system's modularity supports integration with existing hospital information systems, adheres to data governance standards, and is designed to accommodate future XR capabilities.

3.1 Physiotherapist Integration

To ensure clinical safety, the system allows physiotherapists to:

– Approve or reject AI-generated plans.
– Provide inputs and constraints based on their professional judgment.
– Monitor user progress and intervene when necessary.

This hybrid AI-human approach ensures a balance between automation, AI governance, and human expertise.

4   Current Progress

A comprehensive Quality Function Deployment (QFD) analysis was conducted to systematically identify and prioritize user requirements. These insights were then translated into corresponding technical specifications to inform the selection of the most suitable wearable device for data acquisition. Furthermore, a user interface and user experience (UI/UX) prototype was developed using Figma, incorporating interactive components aligned with the specific needs and preferences of both patients and coaches as identified through the QFD process. The avatar design was prototyped as part of this effort. The avatar was also experimented with in an Extended Reality environment using a Meta Quest 3, where the Avatar was placed in an AR setting with passthrough-mode enabled. This mode enables users to chat with the Avatar in a human-like and natural way, without hindering everyday activities. Additionally, the controllers and cameras on the headset provide an understanding of the user's limb positions, enabling the chatbot to guide them through the physical rehabilitation process. This also enables showing the health metrics and additional guidance using floating monitors in the AR space, making the UI more informative and seamlessly integrated with the surroundings as the user keeps moving.

An example dialogue between the chatbot and the user in the current prototype:

AI: How are you feeling today on a scale from 1 to 10 for knee pain?

User: Around 6, especially after walking.

AI: Noted. We'll lower the intensity today and focus on mobility exercises. Ready to begin?

4.1   Evaluation Strategy

Given that the system is currently in its early stages of development, we propose a multi-pronged evaluation framework to assess its functionality, usability, and potential clinical value. First, we will conduct simulated trials using synthetic patient profiles to evaluate the adaptability and responsiveness of the conversational agent under controlled scenarios. These trials will help in identifying whether the system can generate contextually appropriate and varied rehabilitation plans based on different user profiles and input patterns.

In parallel, we will engage physiotherapists in structured expert interviews to review and critique the quality, safety, and clinical relevance of the system's recommendations. Their domain-specific insights will be essential in refining the decision-making logic and ensuring that the outputs align with current rehabilitation best practices.

We also plan to carry out iterative usability testing with representative users, including patients with diabetes and osteoarthritis. These sessions will focus on evaluating the ease-of-use, accessibility, and overall user satisfaction with the avatar-based interface. Feedback from these evaluations will be used to enhance both interaction design and conversational flow.

To quantitatively assess system performance, we will track a range of metrics, including plan adherence rates, dialogue quality (measured using BLEU and ROUGE scores), user-reported engagement and satisfaction levels, and perceived trust in the avatar coach. These evaluation components will collectively inform future iterations of the system and contribute to the evidence base for conversational AI in physical rehabilitation.

5   Conclusion and Future Work

We introduced a hybrid AI-human conversational system aimed at delivering personalized physical rehabilitation for chronic conditions. By combining large language models and clinical supervision, the system adapts to patient needs while ensuring safety and efficacy.

A major innovation is the integration of an avatar-based conversational interface, with a pathway for deployment in XR environments such as AR glasses. This positions the system to evolve alongside the future of immersive health technology.

Future work includes expanding prototype functionality, integrating real-time wearable data, conducting longitudinal user studies, and exploring multi-language and culturally sensitive interfaces. We are also working on different ways to display and interact with the Avatar, ranging from 2D projection on screens, AR on mobile devices, and even holograms. Ultimately, this project seeks to bridge the gap between scalable AI and personalized care.

Acknowledgements. The project is part of work package 3 in the ITEA-TREAT consortium, which aims at transforming healthcare through semantic interoperability and patient self-efficacy. The authors thank the patients, physiotherapists and AI researchers whose feedback is helping shape the development of this system. We would also like to acknowledge Tim Hatkins, who is also a part of the consortium, for his significant contributions to the development and refinement of the system architecture.

References

1. IDF Diabetes Atlas. International Diabetes Federation, 10 edn. (2021)
2. Belpaeme, T., Kennedy, J., Ramachandran, A., Scassellati, B., Tanaka, F.: Social robots for education: a review. Sci. Robot, **3**(21), eaat5954 (2018). https://doi.org/10.1126/scirobotics.aat5954
3. Boltaboyeva, A., Baigarayeva, Z., Imanbek, B., Ozhikenov, K., Getahun, A.J., Aidarova, T., et al.: A review of innovative medical rehabilitation systems with scalable ai-assisted platforms for sensor-based recovery monitoring. Appl. Sci. **15**(12), 6840 (2025). https://doi.org/10.3390/app15126840
4. Devane, N., Behn, N., Marshall, J., Ramachandran, A., Wilson, S., Hilari, K.: The use of virtual reality in the rehabilitation of aphasia: A systematic review. Disabil. Rehabil. **45**(1), 1–20 (2022). https://doi.org/10.1080/09638288.2022.2138573

5. Fitzpatrick, K.K., Darcy, A., Vierhile, M.: Delivering cognitive behavior therapy to young adults with symptoms of depression and anxiety using a fully automated conversational agent (woebot): A randomized controlled trial. JMIR Mental Health **4**(2), e19 (2017). https://doi.org/10.2196/mental.7785
6. Fregna, G., Schincaglia, N., Baroni, A., Straudi, S., Casile, A.: A novel immersive virtual reality environment for the motor rehabilitation of stroke patients: a feasibility study. Front, Robot. AI **9**, 906424 (2022). https://doi.org/10.3389/frobt.2022.906424
7. Griffin, A., Wilson, L., Feinstein, A.B., Bortz, A., Heirich, M.S., Gilkerson, R., et al.: Virtual reality in pain rehabilitation for youth with chronic pain: a pilot feasibility study. JMIR Rehabilitation Assistive Technol. **7**(1), e17957 (2020). https://doi.org/10.2196/22620
8. Lee, M.H.: An intelligent decision support system for stroke rehabilitation assessment. In: Proceedings of the 21st International ACM SIGACCESS Conference on Computers and Accessibility (ASSETS 2019), Pittsburgh, PA, USA,pp. 429–432 (2019). https://doi.org/10.1145/3308561.3356106
9. Mercado, J., Espinosa-Curiel, I., Martínez-Miranda, J.: Embodied conversational agents providing motivational interviewing to improve health-related behaviors: scoping review. J. Med. Internet Res. **25**, e52097 (2023). https://doi.org/10.2196/52097
10. Rainey, J.P., Treu, E.A., Campbell, K.J., Blackburn, B.E., Pelt, C.E., Archibeck, M.J., et al.: Conversational engagement using a short message service chatbot after total joint arthroplasty. Arthroplasty Today **30**, 101484 (2024). https://doi.org/10.1016/j.artd.2024.101484
11. Vinolo Gil, M.J., Gonzalez-Medina, G., Lucena-Anton, D., Perez-Cabezas, V., Ruiz-Molinero, M.D.C., Martin-Valero, R.: Augmented reality in physical therapy: systematic review and meta-analysis. JMIR Serious Games **9**(4), e30985 (2021). https://doi.org/10.2196/30985

Adaptive, Personalized Gamification to Improve Adherence and Retention in Decentralized Study Settings

Janna Herrmann[1,2,3(✉)] ⓘ, Marc Schubhan[1,2] ⓘ, and Maximilian Altmeyer[1] ⓘ

[1] Interaction Experience Group, Saarland University of Applied Sciences (htw saar), Saarbrücken, Germany
{marc.schubhan,maximilian.altmeyer}@htwsaar.de
[2] Saarland Informatics Campus, Saarland University, Saarbrücken, Germany
s9jnherr@stud.uni-saarland.de
[3] ki elements GmbH, Saarbrücken, Germany

Abstract. Decentralized study settings have been proven to increase efficiency and diversity in clinical trials and research studies. However, one of the biggest challenges of decentralized settings is high dropout rates and low adherence to the study protocol. Gamification has shown to be a promising tool to increase motivation and engagement in different settings, but very few tools exist implementing personalization, which is crucial to address the needs of diverse user contexts. In this paper, we present our vision of using gamification to improve adherence and retention in digital, decentralized research studies and clinical trials. We aim to design, implement and evaluate a modular, AI-based gamification engine designed to personalize motivational strategies based on user profiles and context. Its modular, test-agnostic design approach enables the integration across platforms and study types, offering the potential of a reusable infrastructure for adherence support in digital health research.

Keywords: Decentralized Study Designs · Personalized Gamification · Digital Health Engagement · AI in Clinical Interventions · Adherence in Digital Studies · Participant Retention

1 Introduction

Over the last few years, partially driven by the coronavirus pandemic, clinical and research studies have increasingly taken on a decentralized format [8,14]. This approach allows increased efficiency and the inclusion of a broad population that was previously more difficult to reach, as now, fewer on-site visits are necessary. The reduction of travel to study sites and the possibility for more fine-grained interventions opens access to a more diverse participant base, while traditional studies often rely on local participants. Decentralized studies, however, can more easily include individuals from different regions, socioeconomic status or with

C. Fayollas et al. (Eds.): EICS 2025, LNCS 16511, pp. 228–237, 2026.
https://doi.org/10.1007/978-3-032-26051-2_18

specific accessibility needs, which can lead to more representative and robust results.

However, one of the biggest challenges with decentralized studies is maintaining participant adherence and retention, as the reduced direct contact with investigators can make it more difficult to adhere to the prescribed protocols [9]. Resulting high dropout rates can not only reduce statistical power, but can also cause high costs. Therefore, maintaining adherence and minimizing dropouts is an essential goal to ensure the integrity of study results and to optimize the use of resources [12,20]. While adherence and, to some extent, also retention in conventional studies is often ensured by physical visits, decentralized studies must resort to alternative, digital strategies.

Gamification, i.e. the use of game elements in non-game contexts [11], holds great potential in this regard. The integration of game-based, motivational elements into decentralized setups could sustainably improve adherence and retention by adaptively combining meaningful, personalized game elements that motivate participants and promote their steady participation. Game elements such as avatars, narratives, points, badges or levels can increase intrinsic motivation by reinforcing basic psychological needs. Through this, gamification has been shown to have positive effects on motivation, engagement and adherence [13]. In healthcare, gamified systems have been shown to increase long-term motivation among participants by reinforcing feelings of progress and success [10]. If these mechanisms can be transferred to decentralized clinical trials and research studies, they could help to support participants in fulfilling study protocols by increasing the motivation to continue participation. Gamification could also be beneficial as a substitute for the lack of social interaction caused by a lack of direct contact with study staff, playing an important role in maintaining adherence and retention in traditional studies. For example, gamification could be used to promote social interactions between participants or between participants and the study management by using virtual characters and avatars. Current findings in gamification research have shown that both contextual and personal factors play a major role in the success of gamification [1]. Therefore, a dynamic adaptation of the gamification elements used is crucial, as evidence suggests that this can lead to an increase in the effectiveness of gamified interventions [4]. In particular, this means that the specific context and the interpersonal differences of the participants must be taken into account when selecting and implementing game elements.

In this position paper, we present our vision to improve adherence, retention, and user experience (UX) in digital, decentralized studies in clinical research through personalized gamification. Our proposed approach is in the conceptual phase, but the design and evaluation plans outlined in this paper lay the groundwork for future development and pilot testing. The topic is highly relevant and timely: The COVID-19 pandemic has catalyzed a shift from site-based protocols to decentralized studies and clinical trials [8,14], but low adherence is a major issue [9]. Concurrently, the gamification literature is progressing beyond "one-size-fits-all" toward personalized approaches, with recent findings indicating more effective outcomes [4]. Thus, our proposed approach—utilizing AI-based recommendations to tailor gamification elements in order to enhance adherence in clinical trials—is a relevant and innovative contribution that aligns with current trends in both digital health and personalized user engagement. More specifically, we aim to design, implement and evaluate a modular, AI-based gamification engine that enables context- and user-dependent, dynamic adaptation of used gamification elements. The modularity of this engine is at the forefront of the implementation, such that the elements can be embedded into existing platforms for planning, managing and conducting decentralized studies. By strengthening UX, adherence and retention in decentralized research, the vision supports the generation of higher-quality datasets, which can in the long run enable earlier detection of health changes, more robust preventive insights, and improved health outcomes across diverse populations.

2   Position and Proposed Approach

Current approaches for increasing adherence and retention often rely on the usage of reminders, incentives or the general idea of user-friendly interfaces resulting in static and generic systems that insufficiently cater for the diverse user needs and contexts in decentralized study settings [6,7]. As stated by Carlier et al. [6], compared to other domains, gamification in healthcare requires the integration of expert knowledge, dynamic user profiling and personalization. Also, the lack of an overall standardized design process emphasizes the need for reusable algorithms. Key factors identified to contribute to the success of adaptive personalized gamification in healthcare include goal-setting, feedback mechanisms, social interaction, and reward systems tailored to individual preferences [21]. In this paper, we challenge the idea of the one-size-fits-all approach and propose our vision of an adaptive, personalized gamification system consisting of a modular gamification engine that personalizes game elements and mechanics based on user profiles and context. In the long term, this adaptive module should enable sustainable, ethical, user-centered engagement across diverse populations.

Several aspects distinguish our approach from conventional gamification strategies in clinical research. First, our approach will be modular and test-agnostic to ensure that it can be flexibly integrated across platforms and study types rather than being tailored to a single platform, a certain study protocol, or a specific disease. Second, our vision combines AI-based personalization with rule-based safeguards to create a robust, hybrid system that is adaptive to individual user preferences while being mindful of clinical, ethical, and accessibility considerations. Moreover, by exploring the integration of user-type models like HEXAD [22] into the personalization, we are able to leverage a systematic understanding of motivational diversity which has already been proven to be successful in other contexts [2,3]. Finally, our approach explicitly addresses clinical validity throughout by prioritizing transparent personalization and considering how game elements might interfere with it. Together, these aspects make our vision uniquely suited to the demands of decentralized clinical research.

To demonstrate how our vision translates into practice, we outline two illustrative scenarios:

Scenario 1: In Alzheimer's disease trials, high screen failure rates are common [18]. One way to address this is by remotely prescreening participants at home to assess their eligibility. Gamification could make these assessments more engaging and improve compliance. However, older participants might struggle with high-speed or complex tasks. Here, our system would avoid time-limited elements or complex narratives that could create stress or cognitive overload. Instead, participants would benefit from achievement-based badges, gentle rewards, and simple narratives (e.g., "Help us learn more about your memory").

Scenario 2: Depression trials often have small sample sizes [17], making adherence critical. Gamification could help increase adherence, but competitive features like leaderboards might worsen feelings of inadequacy and hopelessness in participants with depression. In contrast, healthy controls might still benefit from these elements. Therefore, the choice of game elements is critical, and the system should adapt to individual users.

The theoretical basis of our approach is grounded on the assumption, that adherence and retention are dynamic behaviors influenced by many factors including motivation, context and UX. Therefore, instead of using static gamification elements, the use of adaptive user-specific factors is crucial to maintain user engagement over time. To address this, we envision an AI-based dynamic gamification engine, that draws from a set of predefined gamification elements, such as progress tracking, challenges, narratives, score systems etc. based on factors like engagement history, study context, demographics (e.g. age, gender), diagnosis, and user preferences. The selected elements will be test-agnostic to ensure the gamification layer can be used across various study types and participant populations. While the exact elements of this gamification library will be developed interactively throughout the development, the initial selection and strategies for personalization will be based on both rule-based systems as well as data-driven models to be able to find a balance between engagement and protocol fidelity/clinical validity.

A key component for building the data-driven pillar of the personalization strategy is the development of a crowdsourced dataset to support the training and evaluation of the AI-based gamification model. The crowdsourcing study aims to empirically identify relationships between user characteristics and gamification preferences or effectiveness, specifically looking at UX, adherence and retention. The study will be designed in a way that allows participants to rate, experience, or rank game elements individually across different task-based scenarios. This ensures that personalization strategies are grounded in user responses to real, representative game dynamics.

To achieve this, we plan to collect data across three categories of personalization-relevant factors: Starting with demographic data (e.g. age, gender, education) as it has a significant influence on how individuals interact with and perceive gamified systems [5]. Secondly, we will focus on player types (e.g. HEXAD [15,22]), as they have been shown to correlate with specific game elements [16], which allows a personalization based on individual preferences inferred by these types, such as an increased need for achievement or social interaction among others. Lastly, we will incorporate medical information (e.g. diagnostic data or clinical scales) as this allows us to consider the game elements' impact and suitability to the individual health condition [19]. While the first two categories can be assesses via questionnaires, we aim to assess the medical information via remote cognitive screening tools, such as digital speech biomarker, or clinical assessment scores.

Before launching the crowdsourcing study, we will conduct an internal selection process to define a diverse and theory-informed set of gamification elements to include in the study. The aim is to curate a balanced set of elements that vary in motivational type (e.g., achievement-based, social, narrative-driven) to explore which combinations are most effective for which user types. This selection will be based on:

1. Prior research in health gamification (e.g. [10,13])
2. Commonly used motivational mechanics
3. Alignment with psychological theories of motivation (e.g., Self-Determination Theory)
4. Compatibility with decentralized study workflows

The second pillar of the personalization and gamification strategy will be rule-based to support initial deployments and address ethical, accessibility or clinical validity concerns. For example, the use of competitive elements may not be appropriate for certain clinical populations, or the use of certain visual or narrative features may result in emotional distress or cognitive overload for specific groups of participants. The foundation for the rules will be the knowledge of domain experts as well as a literature review. This rule-based layer will allow the system to make safe, conservative decisions when personalization data is limited or when exclusion criteria are triggered. Each exclusion rule should be linked to at least one of the following non-exhaustive list that may be extended as the system evolves:

1. **Preservation of Clinical Validity** (e.g. "Avoid game elements resulting in a delayed response in time-sensitive assessments.")
2. **Psychological Safety** (e.g. "Avoid emotionally intensive or negative narrative elements for participants with or at risk of depression, anxiety etc.")
3. **Accessibility and Usability** (e.g. "Avoid game elements relying on heavy reading for participants with reduced cognitive abilities")
4. **Protocol Compatibility** (e.g. "Avoid game elements that alter with the task flow if there is a fixed protocol/order of the tasks.")

Ultimately, the combination of data-driven and rule-based personalization aims to allow a balance between user engagement, accessibility, ethical responsibility, and protocol fidelity.

3   Toward Evaluation and Impact

Through this position paper, we outline the planned evaluation strategy for our vision as a foundation to foster discussion:

The system should be first evaluated in a non-clinical study to evaluate feasibility, technical functionality, UX in general and to look at first evidence of personalization effects. A following clinical pilot study should evaluate the effect in a real-word decentralized study setting to assess its translational value and evidence needed for further adaptation. It allows to draw conclusions about whether the system safely increased adherence and retention while preserving protocol integrity as well as measuring acceptability to both participants and clinicians. Both studies will be randomized to allow the isolation of the effect of personalized gamification from other factors.

We consider the following operationalization to measure outcomes and success in the evaluation: **Adherence** may be measured as how well participants follow the expected behavior within a burst or time window of the study, e.g. the number of completed scheduled tasks (5 out of 6 planned assessments) and/or the task completions on time. **Retention** measures whether participants remained in the study over time, which can be operationalized by the percentage of participants completing the full protocol. The **UX** will be quantitatively assessed using established, validated psychometric tests and supplemented by qualitative methods. To assess whether the **validity of the underlying task or measure** is preserved, a control group will be established and the mean scores and variability will be compared. Additional approaches could be the assessment of the test-retest reliability of the gamified approach to measure, whether the results are stable over time even when gamification elements change or adapt. High correlation between the sessions would suggest that the measurement preserved its reliability.

4   Key Considerations for Personalized Gamification in Health Research

Even if this approach sounds rather straightforward, several key considerations need to be taken into account, already during the design process. In the following, we will highlight key technical, clinical, and ethical considerations aiming to foster the discussion around open questions in adaptive engagement strategies in health research.

4.1   Clinical Considerations

First and foremost, the most important consideration is to ensure that the gamification elements do not interfere with the clinical validity of the study and its assessment. The selection of gamification elements must be carefully considered and validated to ensure the elements do not interfere with dependent variables, e.g. by distracting users or causing fatigue. The core clinical tasks must stay unchanged, even as the motivational mechanisms around it are dynamically adapted. Another clinical consideration is the personalization of UX. Although the experience is tailored to users, the overall consistency of the data collection across participants must not be tempered, and adaptive gamification must remain in compliance with the study protocol and ethical oversight. Last but not least, it must also be considered how clinicians interact with the gamification system and how they can be provided with the necessary transparency and control over it.

4.2   Technical Considerations

Looking at the technical considerations for the engine, the most prominent one is the question, which data points can be reliably used for personalization (e.g. demographics, diagnosis, study context). In most study settings, general demographic information and potentially the diagnosis as well as the study context itself might be known, but still could be sparse or incomplete. Given that, one key consideration for the design of the adaptive gamification engine is to identify which features are predictive of motivational preferences without creating excessive burden on participants, study teams or introducing additional privacy risks. An additional consideration is given by the general cold-start problem in personalization. In early stages of the interaction, the system still has limited knowledge and with that, personalization will be less effective. To ensure a more meaningful personalization early on the system could include questionnaires (e.g. player-type questionnaires, such as Hexad [15]) or default motivational profiles based on population-level data. However, both require validation in clinical contexts. The choice of personalization logic is crucial as well and comes with an inherent trade-off between transparency given by rule-based approaches and the potential benefit of more nuanced adaptations given by machine learning approaches which, in contrast, are less transparent and might be more difficult to evaluate. Therefore, we aim to explore hybrid strategies. Moreover, the frequency of adaptation of the systems must be considered as well. While real-time personalization might enhance responsiveness, it also comes with the disadvantage of increased computational load and potential user distraction as well as increased intransparency for the study itself. Alternatively, periodic, predefined update cycles might be more feasible and preferred in the clinical system to make the user flow more predictable.

4.3   User Experience and Acceptability Considerations

Besides the discussed clinical validity, gamification might also affect the participants' burden and their perceived credibility of the trial. It must be considered how to ensure that the system is still adequate regarding the credibility and trustworthiness needed in clinical study settings, while at the same time being motivating and game-like. An additional important consideration to make is that personalization needs to be culturally and demographically aware to prevent unintentional exclusion or alienation of participants due to their cultural background, demographics (e.g. age) or digital literacy. Non-aware gamification systems might use styles, symbols or language that does not resonate with some participants leading to disengagement, reduced trust or unequal adherence rates across different user subgroups.

4.4   Ethical Considerations

Ethical considerations are crucial when it comes to clinical settings. Transparency of the system is one of the key considerations and raises the question how much users should (or must) know about personalization to ensure ethical considerations are met while still being motivational for the user without breaking immersion. Manipulative design and over-personalization must be avoided and it must be ensured, that personalization strategies do not favor certain groups. With that, the protection of participant data and how it is used for personalization must also be considered. Any user studies conducted as part of the development or evaluation of this system will undergo review and approval by institutional ethics review boards to ensure compliance with ethical guidelines and regulatory standards.

5   Conclusion and Future Directions

This position paper states that in order to improve adherence and retention in decentralized studies, a shift from static engagement mechanisms toward adaptive, personalized solutions is needed. Gamification has shown to be a promising tool to increase motivation and engagement in different settings, but very few tools implementing personalization exist which is crucial to address the needs of diverse user contexts. We propose a novel approach by aiming to develop an AI-based modular gamification engine designed to personalize motivational strategies based on user traits and context. Its modular, test-agnostic design approach enables cross-platform integration offering the potential of a reusable infrastructure for adherence support in digital health research. In addition, the success of our vision and similar approaches will depend on the level of attention to ethical, clinical and user-centered design considerations. We aim to explore whether our hypothesis holds true, i.e. personalization improves adherence and user experience without compromising clinical validity in decentralized settings. At present, this work represents a conceptual vision that will guide future implementation and evaluation in real-world clinical studies. Future work will involve the selection of the personalization strategies, gamification elements and the validation in diverse populations with regards to the effectiveness of improving adherence, retention and UX while maintaining clinical validity. If successful, our approach could grant important insights for enabling more inclusive, representative digital health research, particularly in areas where high granularity or long-monitoring with high user engagement is essential.

References

1. Altmeyer, M.: Understanding personal and contextual factors to increase motivation in gamified systems (2022). https://doi.org/10.22028/D291-36248
2. Altmeyer, M., Lessel, P., Jantwal, S., Muller, L., Daiber, F., Krüger, A.: Potential and effects of personalizing gameful fitness applications using behavior change intentions and hexad user types. User Model. User-Adapted Interact. **31**, 675–712 (2021). https://doi.org/10.22028/D291-33887
3. Altmeyer, M., Schubhan, M., Lessel, P., Muller, L., Krüger, A.: Using hexad user types to select suitable gamification elements to encourage healthy eating. In: Extended Abstracts of CHI 2020 (2020). https://doi.org/10.1145/3334480.3383011
4. Altmeyer, M., Zenuni, B., Spelt, H., Jegen, T., Lessel, P., Krüger, A.: Do Hexad user types matter? effects of (non-) personalized gamification on task performance and user experience. In: Proceedings of the ACM on Human-Computer Interaction vol. 6(CHI PLAY), pp. 1–27 (2022). https://doi.org/10.1145/354949
5. Aydin, G.: Effect of demographics on use intention of gamified systems. Int. J. Technol. Human Interact. **14**(1), 1–21 (2018). https://doi.org/10.4018/IJTHI.2018010101
6. Carlier, S., De Backere, F., De Turck, F.: Personalised serious games and gamification in healthcare: Survey and future research directions (2024). https://doi.org/10.48550/arXiv.2411.18500
7. Carreño, A.M.: A framework for agile design of personalized gamification services (2018)
8. Cummins, M.R., Burr, J., Young, L., Yeatts, S.D., Ecklund, D.J., Bunnell, B.E., et al.: Decentralized research technology use in multicenter clinical research studies based at U.S. academic research centers. J. Clin. Trans. Sci. (2023). https://doi.org/10.1017/cts.2023.678
9. Dahlhausen, F., Zinner, M., Bieske, L., Ehlers, J.P., Boehme, P., Fehring, L.: There's an app for that, but nobody's using it: insights on improving patient access and adherence to digital therapeutics in Germany. DIGITAL HEALTH **8** (2022). https://doi.org/10.1177/20552076221104672
10. De Croon, R., Geuens, J., Verbert, K., Vanden Abeele, V.: A systematic review of the effect of gamification on adherence across disciplines. In: Fang, X. (ed.) HCII 2021. LNCS, vol. 12789, pp. 168–184. Springer, Cham (2021). https://doi.org/10.1007/978-3-030-77277-2_14
11. Deterding, S., Dixon, D., Khaled, R., Nacke, L.: From game design elements to gamefulness: defining "gamification". In: Proceedings of the 15th International Academic MindTrek Conference (2011). https://doi.org/10.1145/2181037.218104
12. Forbes, A., Keleher, M.R., Venditto, M., DiBiasi, F.: Assessing patient adherence to and engagement with digital interventions for depression in clinical trials: systematic literature review. J. Med. Internet Res. **25** (2023). https://doi.org/10.2196/43727
13. Hamari, J., Koivisto, J., Sarsa, H.: Does gamification work?–a literature review of empirical studies on gamification. In: 2014 47th Hawaii International Conference on System Sciences (2014). https://doi.org/10.1109/HICSS.2014.377
14. Hanley Jr, D.F., Bernard, G.R., Wilkins, C.H., Selker, H.P., Dwyer, J.P., Dean, J.M., et al.: Decentralized clinical trials in the trial innovation network: Value, strategies, and lessons learned. J. Clin. Trans. Sci. **7** (2023). https://doi.org/10.1017/cts.2023.597

15. Krath, J., Altmeyer, M., Tondello, G.F., Nacke, L.E.: Hexad-12: Developing and validating a short version of the gamification user types hexad scale. In: Proceedings of the 2023 CHI Conference on Human Factors in Computing Systems, pp. 1–18 (2023). https://doi.org/10.1145/3544548.3580968
16. Krath, J., von Korflesch, H.F.O.: Player types and game element preferences: investigating the relationship with the gamification user types HEXAD scale. In: Fang, X. (ed.) HCII 2021. LNCS, vol. 12789, pp. 219–238. Springer, Cham (2021). https://doi.org/10.1007/978-3-030-77277-2_18
17. Liang, J., He, P., Wu, H., Xu, X., Ji, C.: Characteristics of depression clinical trials registered on ClinicalTrials.gov. Inter. J. General Med. **15**, 8787 (2022). https://doi.org/10.2147/IJGM.S394143
18. Malzbender, K., Lavin-Mena, L., Hughes, L., Bose, N., Goldman, D., Patel, D.: Key barriers to clinical trials for Alzheimer's disease. Tech. rep., Leonard D Schaaeffer Center for Health Policy & Economics (2020)
19. Pavarini, G., et al.: Ethical implications of digital gaming interventions for mental health: systematic review and critical appraisal (2024). https://doi.org/10.31219/osf.io/a6pvx
20. Ridho, A., Alfian, S.D., van Boven, J.F., Levita, J., Yalcin, E.A., Le, L., et al.: Digital health technologies to improve medication adherence and treatment outcomes in patients with tuberculosis: Systematic review of randomized controlled trials. J. Med. Internet Res. **24**(2) (2022). https://doi.org/10.2196/33062
21. Tera, T., Kelvin, K., Tobiloba, B.: Evaluating gamified interventions in healthcare practice (2024). https://doi.org/10.31234/osf.io/z94eh
22. Tondello, G.F., Wehbe, R.R., Diamond, L., Busch, M., Marczewski, A., Nacke, L.E.: The gamification user types hexad scale. In: Proceedings of the 2016 Annual Symposium on Computer-Human Interaction in Play, pp. 229–243 (2016). https://doi.org/10.1145/2967934.2968082

Author Index

C. Fayollas et al. (Eds.): EICS 2025, LNCS 16511, pp. 239–240, 2026.
https://doi.org/10.1007/978-3-032-26051-2